Electrochemical Methods in Archaeometry, Conservation and Restoration

Monographs in Electrochemistry

Surprisingly, a large number of important topics in electrochemistry is not covered by up-to-date monographs and series on the market, some topics are even not covered at all. The series *Monographs in Electrochemistry* fills this gap by publishing in-depth monographs written by experienced and distinguished electrochemists, covering both theory and applications. The focus is set on existing as well as emerging methods for researchers, engineers, and practitioners active in the many and often interdisciplinary fields, where electrochemistry plays a key role. These fields will range – among others – from analytical and environmental sciences to sensors, materials sciences and biochemical research.

Information about published and forthcoming volumes is available at
http://www.springer.com/series/7386

Series Editor: Fritz Scholz, University of Greifswald, Germany

Antonio Doménech-Carbó
María Teresa Doménech-Carbó
Virginia Costa

Electrochemical Methods in Archaeometry, Conservation and Restoration

Springer

Dr. Antonio Doménech-Carbó
Universitat València
Fac. Química
Depto. Química Analítica
C/Dr. Moliner, 50
46100 Burjasot Valencia
Spain
antonio.domenech@uv.es

Prof. Dr. María Teresa Doménech-Carbó
Universitat Politècnica de València
Inst. Restauració del Patrimoni
Depto. Conservació i
Restauració de Bens Culturals
Camí de Vera, s/n
46022 València
Spain
tdomenec@crbc.upv.es

Prof. Dr. Virginia Costa
15 rue Jules Hetzel
92190 Meudon
France
virginia.costa@gmail.com

ISBN 978-3-540-92867-6 e-ISBN 978-3-540-92868-3

DOI 10.1007/978-3-540-92868-3

Library of Congress Control Number: 2008942070

Cover design: WMXDesign, Heidelberg

Printed on acid-free paper

9 8 7 6 5 4 3 2 1

springer.com

Preface

Electrochemical systems—e.g., batteries, capacitors, and fuel cells—are an integral part of modern technology. Electrochemical techniques, especially potentiometry and voltammetry, are indispensable for state-of-the-art analysis, and also for fundamental studies of the properties of solution species and solid phases and materials. Last, but not least, electrochemical concepts for understanding charge transfer reactions entered the fields of biochemistry and biophysics. Indeed, the last century is characterized by a constant expanding of applications into more and more fields of science and technology. Among these new applications are also archeometry, conservation, and restoration. Electrochemistry is well-suited for these destinations because it can be used to investigate metals and alloys, and it can be used to analyze solid phase systems. Electrochemistry makes thermodynamical and kinetic-based information accessible on redox systems that are frequently constituents of archaeological and historic objects. The authors of this monograph are three leading scientists who are very well-known for their seminal contributions to the use of electrochemical techniques for analytical characterization of such objects, as well as for their restoration. The authors are equally well-acquainted with all the other techniques, be it spectroscopic, diffraction, chromatographic, or microscopic techniques, etc. This allows them to present an unbiased view on electrochemical techniques. This monograph is the result of extensive experience with electrochemistry, and the analysis and restoration of objects of cultural heritage. Thus, I am very thankful to siblings Teresa and Antonio Doménech-Carbó from Spain, and to Virginia Costa from France for having agreed to co-write this monograph, which will be beneficial for two rather different communities: a) electrochemists and electroanalysts, who may learn a lot about the fruitful application of their techniques for the analytical characterization of historic objects, and also for conservation and restoration, and b) scientists working in laboratories of conservation, restoration, and analysis of historic objects, who can learn a lot about the high potential of electrochemistry in their work. The fact that the monograph had to be written for these two scientific communities posed serious problems because normally, the scientists of these two communities will not know much about each others' science. I think that the authors of the monograph have found a very good way to respond to the needs of all readers for understanding this very interdisciplinary topic, and I hope that the book will be beneficial for the broad audience it addresses.

September 2008

Fritz Scholz
Editor of the series *Monographs in Electrochemistry*

Contents

List of Acronyms

AA:	activation analysis
AAS:	atomic absorption spectroscopy
AE:	acoustic emission
AES:	Auger electron spectroscopy
AFM:	atomic force microscopy
AMS:	accelerator mass spectrometer
ATR:	attenuated total reflection
CCD:	charge coupled device
CE:	capillary electrophoresis
CLSM:	confocal laser scanning microscopy
CPE:	carbon paste electrodes
CT:	computed X-ray tomography
CV:	cyclic voltammetry
DC:	direct current
DPMS:	direct pyrolysis mass spectrometry
DPV:	differential pulse voltammetry
DRIFT:	diffuse reflection Fourier-transform infrared spectroscopy
DSC:	differential scanning calorimetry
DTA:	differential thermal analysis
DTMS:	direct temperature-resolved mass spectrometry
EDX (EDS):	energy dispersive x-ray microanalysis
EELS:	electron-energy-loss spectroscopy
ELISA:	enzyme-linked immunosorbent assays
EPXMA:	electron probe x-ray microanalysis
EPMA:	electron probe microanalysis
EPR:	electron paramagnetic resonance spectroscopy
ESEM:	environmental scanning electron microscopy
ESR:	electron spin resonance spectroscopy
FAB-MS:	fast-atom bombardment mass spectrometry
FID:	flame ionization detector
FLIM:	fluorescence lifetime imaging
FTIR:	Fourier transform infrared spectroscopy
FTIR-PAS:	FTIR photoacoustic spectroscopy

FTO: fluorine-doped tinoxide
GC: gas chromatography
GCE: glassy carbon electrode
GC-MS: gas chromatography- mass spectrometry
HINDT: holographic interferometry nondestructive testing
HPLC: high performance liquid chromatography
IC: ion chromatography
ICP-AES: inductively coupled plasma-atomic emission spectroscopy
ICP-MS: inductively coupled plasma-mass spectrometry
ICP-SMS: inductively coupled plasma sector field mass spectrometry
ICP-TOF-MS: inductively coupled plasma time-of-flight mass spectrometry
IFM: immunofluorescence microscopy
LA-ICPMS: laser ablation inductively coupled plasma-mass spectrometry
LC: liquid chromatography
LDI: laser desorption/ionisation
LDMS: laser induced desorption mass spectrometry
LIBS: laser induced breakdown spectroscopy
LIF: laser-induced fluorescence
LM: light microscopy
LSV: linear potential scan
MALDI: matrix-assisted desorption/ionisation
μ-SRXRD: micro-x-ray diffraction
µ-XRF: microscopic XRF
μ-SRXRF: μ-XRF spectrometry synchrotron-based
μ-XANES: micro-x-ray absorption near-edge structure spectroscopy
MC-ICP-MS: multiple collector inductively coupled plasma mass spectrometry
MRI: magnetic resonance imaging
MS: mass spectrometry
MS-MS: tandem mass spectrometry
NAA: neutron activation analysis
nanoLC-nanoESI-Q-qTOF-MS-MS: nano liquid chromatography- electrospray ionization-quadrupole time-of-flight tandem mass spectrometry
NMR: nuclear magnetic resonance
OSL: optically stimulated luminescence
OSP: optically stimulated phosphorescence
PAA: proton activation analysis
PAS: photoacoustic spectroscopy
PC: paper chromatography
PDA: photodiode-array
PFE: polymer film electrode
PIGE: particle-induced gamma-ray emission spectroscopy
PIXE: proton-induced x-ray emission spectroscopy
PTTL: phototransferred thermoluminiscence

Py-GC-MS: pyrolysis gas chromatography mass spectrometry
Q-ICPMS: quadrupole inductively coupled plasma mass spectrometry
RBS: Rutherford backscattering spectrometry
SAM: scanning Auger microscopy
SCE: saturated calomel (reference) electrode
SEC: size-exclusion chromatography
SEM: scanning electron microscopy
SCLF: single crystal laser fusion
SIMS: Secondary ion mass spectrometry
SIPS: sputter-induced optical spectrometry
SLDV: laser Doppler vibrometry
SPI: speckle pattern interferometry
SPS: speckle pattern shearography
SQWV: square wave voltammetry
STEM: scanning transmission electron microscopy
TEM: transmission electron microscopy
TG: thermogravimetric analysis
TIMS: thermal ionisation mass spetrometry
TL: thermoluminiscence
TLC: thin-layer chromatography
TRM: thermoremanence
XAS: μ-x-ray absorption spectroscopy
XPS: x-ray photoelectron spectroscopy
XRD: x-ray diffraction spectrometry
XRF: x-ray fluorescence spectrometry

Introduction

Scientific examination of archaeological pieces and works of art is undoubtedly a necessary task for archaeometry, conservation and preservation/restoration sciences. Although essentially focused on metal corrosion problems, electrochemistry was one of the early applied scientific methodologies in such fields, in both its analytical and conservative/restorative aspects. Over the last few decades, the scope of electrochemical methods' ability to interact with archaeometry, conservation and restoration has been significantly extended, by virtue of the application of new approaches—in particular, the voltammetry of microparticles.

The current monograph is devoted to presenting the state of the art with regard to the intersection between electrochemistry, archaeometry, conservation, and related scientific fields. The intention in writing this book was to make electrochemical methods accessible to a reader interested in the study of cultural heritage, but not necessarily familiarized with electrochemistry. Conversely, this book is also devoted to electrochemists interested in the possibilities of their scientific branch in the context of studies of cultural goods. In this sense, although the book summarizes the content of more than 300 publications (listed in the bibliography), the text concentrates on branches of electrochemistry directly related with studies on archaeological pieces and/or works of art.

The book has been structured into roughly three parts. First (Chap. 1), an overview of analytical methods applied in the study of cultural goods is presented to situate electrochemical methods in their analytical context. The second part contains voltammetric methods devoted to the identification (Chap. 2), speciation (Chap. 3), and quantitation (Chap. 4) of microsample components from works of art and/or cultural and archaeological pieces. The third part of the book presents selected examples of the deterioration of metal artifacts, outlining aspects peculiar to the cultural heritage conservation field (Chap. 5), and describes historic and current issues regarding electrochemical techniques used in restoration treatments and preventive conservation (Chap. 6).

We would like to express our appreciation and thanks to Fritz Scholz for his accurate and inspired revision of the text, and the support of our colleagues—Francisco Bosch, José Vicente Gimeno, Sinforiano Sánchez, and Rufino Mateo from the University of Valencia; María Luisa Vázquez, Isabel Martínez, and Dolores Julia Yusá from the Polytechnical University of Valencia; and our collaborators Laura Osete,

Juana De la Cruz, María José Casas, Montserrat Moyá, Juan Peris, Julia Ciarrocchi, Vanja Cialei, Marina Calisti and Vincenzo Maiolo.

Antonio Doménech-Carbó
María Teresa Doménech-Carbó
Virginia Costa

Chapter 1
Application of Instrumental Methods in the Analysis of Historic, Artistic and Archaeological Objects

1.1 Importance of Scientific Examination for Archaeometry, Conservation and Restoration

Preservation of cultural goods is an important and rewarding task of modern societies. They are a vital source of inspiration and reflect the culture and history of the past and present. This valuable asset is the basis for future cultures and therefore, is one of the main legacies to be passed on to future generations. For this purpose, conservators, curators, art historians, and scientists combine their efforts, making it now a pluridisciplinary activity.

Cultural goods are particularly rich and diverse, as they are comprised of a great variety of materials (and often made from combinations thereof), and they are of vastly different sizes ranging from archaeological or historical sites, to monuments, and to objects of fine craft or art. Although their significance stems from the transmitted historical, cultural, or figurative messages, their conservation and perpetuation in time depends on their materials. The *Science of Conservation* has been developed for this reason, and is devoted to the scientific study of the objects and the procedures that assure the safeguarding of cultural goods.

Application of the physical and chemical sciences to tackle the problems and questions of archaeology, history, and the conservation of heritage dates back to the 18th century. The first specialized laboratory dedicated to this type of work was established in 1888: the *Chemisches Labor der Königlichen Museen zu Berlin* (Chemical Laboratory of the Royal Museums of Berlin, now the *Rathgen-Forschungslabor*). This laboratory was set up based on the idea that the scientific approach to cultural goods is always ancillary to the approach made by the art historian and the conservator. Currently, scientific disciplines play an essential role in the material characterization of art objects (Scheme 1.1). For example, the dating of archaeological remains is based on instrumental techniques. Characterization of the artistic techniques and technologies of production from the analytical data related to the chemical composition and morphology of the object often allows a clear ascription of the studied object to a geographical region, as well as an elucidation of the date of manufacture. Authentication is sometimes carried out based on analytical data (i.e., identification of a pigment used in a certain historic period or recognition

A. Doménech-Carbó et al., *Electrochemical Methods in Archaeometry, Conservation and Restoration*, Monographs in Electrochemistry,
DOI 10.1007/978-3-540-92868-3_1, © Springer-Verlag Berlin Heidelberg 2009

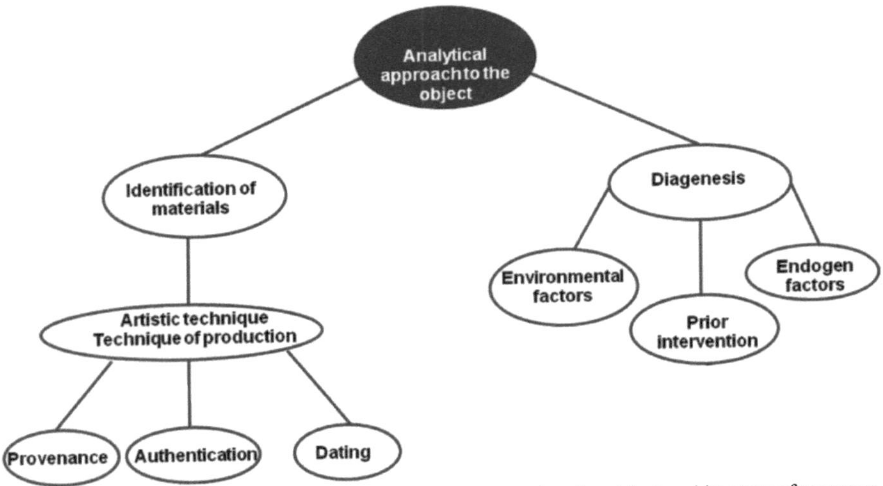

Scheme 1.1 Specific issues related to the materials composing the object and its state of preservation directly derived from analytical data

of a specific artistic technique). Analytical data are essential for determining the state of conservation of the object, as well as the causes and mechanisms of its deterioration. Three principal sources of deterioration are investigated: environmental factors resulting in mechanical, chemical, or biological alterations; endogen causes due to incompatibility of materials present in the object; and the ability of the object to undergo autodegradation or alterations due to old restorations. Scientific examination of the object provides a complete picture of the type of damage it exhibits: occurrence of changes in morphology and/or composition, formation of corrosion products, lixiviation of materials, etc.

Subsequently, based on this fundamental knowledge, it is further helpful to define, develop, and evaluate conservation concepts, materials, measures, methods, and techniques of intervention (Scheme 1.2). Analytical control of the intervention

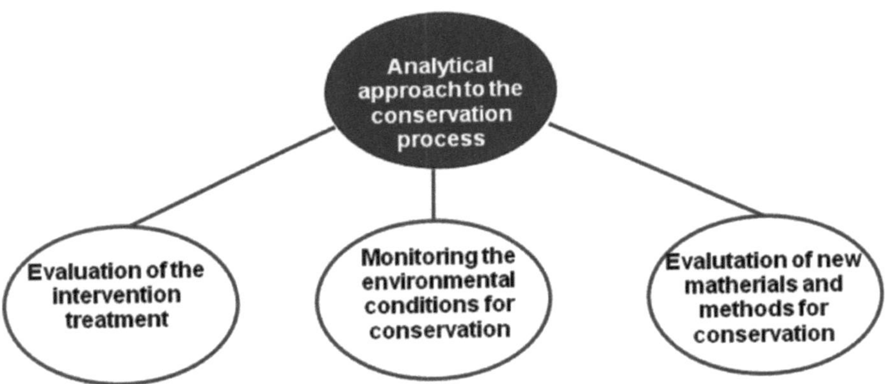

Scheme 1.2 Specific issues related to the conservation and preservation treatments directly derived from analytical data

treatment is essential for assuring that the operations included in the conservation treatment are performed correctly. Monitoring of the environment is another fundamental task for guaranteeing the future conservation of the object under optimal conditions. Control of temperature, relative humidity, and pollutants in museums (exhibition and storage rooms), monuments, and buildings of interest, as well as at archaeological sites, are indispensable activities carried out during the intervention of the object and beyond. These are good examples of infield applications of analytical techniques. In parallel, research in the science of conservation is carried out in the laboratory, which is focused on the development of new, more selective and sensitive analytical methods, as well as new materials and methods for conservation.

1.2 Information Provided by the Analytical Research

Analytical techniques offer a great deal of information that is relevant to cultural goods and their preservation conditions. Analytical data provide connections among causes and effects, and they are the basis for establishing theoretical models that describe the alteration processes exhibited by the object. In this section, let us consider the analytical information according to its source: the object, the environment, and the conservation process.

1.2.1 Analytical Information Obtained from the Object

Analytical data from the object can be grouped as follows:

(a) Morphological information. Technical examination of the object provides information concerning the size, shape, and method of manufacture, as well as the presence of damage. The examination of the object can be performed with respect to the bulk, to a specific part of its surface, or with respect to a microsample prepared as a thin section or a cross section.

Morphological changes observed in the bulk or on the surface of the object are associated with the damages inflicted upon the object: debris, dust, superficial deposits, crusts, cracks, pores, fissures, fractures, laminations, lixiviations, spots, efflorescences, etc. (Figs. 1.1 and 1.2).

Examination of thin sections prepared from microsamples of organic materials such as wood, parchment, textile, paper, ivory, horn, or leather enables the recognition of anatomical and histological features characteristic of the type of organic material or botanical specie.

Examination of cross sections (in particular) of samples extracted from polychrome objects provides the complete sequence of pictorial strata present in the object, as well as the possible infiltrations or corrosion crusts formed as consequence of the alteration processes (Fig. 1.3). The distribution of pictorial layers is also essential for establishing the artistic technique used by the artist.

Fig. 1.1 Cracks and pores formed on the surface of a glazed tile (Manises, 18th century) from the floor of the balconies of the Basílica de la Virgen de los Desamparados de Valencia, Spain

Fig. 1.2 Detail of the bleaching formed on the surface of an oil painting. Examination with stereomicroscope denotes the formation of microcrystals on the surface of the paint layer, which are responsible for the noticeable change of the visual appearance of the painting

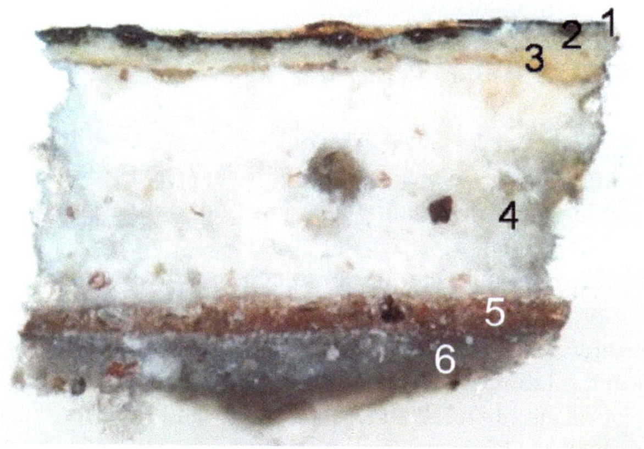

Fig. 1.3 Cross section of an ancient wall painting illustrating the strata distribution: on the top a thin translucent protective coating can be recognized, (1); green paint layer, (2); white paint layer, (3); and thick layer of ground, (4). Underneath all that an original clayey paint layer (5) and its corresponding ground (6) can be recognized

(b) Physical information. Physical, mechanical, and optical properties of the object characterize the material behavior. Consequently, the changes in their original values indicate that some alteration process took place. Among the properties commonly determining the mechanical behavior of the object include density, Young's modulus, ultimate tensile strength, ultimate compressive strength, and ultimate flexion strength. Other physical properties that determine the behavior of the solid object are those related to the porous structure of the material that conforms the object—namely, saturation coefficient, water vapor conductivity coefficient, water absorption coefficient, and permeability.

Optical properties such as color, pleochroism, refractive index, and birefringence, among others, discovered on thin sections when they are examined by means of a petrographic microscope, are essential for characterizing rocks (sculpture and architectural materials). They also provide useful information for characterizing pigments, ceramics, glass and glazes, plasters, metals, and slags, as well as for recognizing alteration processes.

(c) Chemical information. A wealth of qualitative and quantitative chemical information can be obtained from the materials and alteration products formed on the object. Depending on the analytical technique used, that information can be very diverse. Elemental composition is a usual type of data. Structural information can also be available, including recognition of functional groups or of the complete molecular structure, mineralogical distribution, degree of crystallinity, and cell parameters. Presence of isomeric species is frequently detected in organic materials. Electrochemical techniques enable identification of electroactive species, and make speciation studies of the examined materials possible. Calorimetric data are available from thermoanalytical techniques.

In chromatographic techniques, qualitative identification of organic binders requires quantification of species obtained from the original polymeric materials: amino acids from proteins, fatty acids from drying oils, or monosaccharides in plant gums.

Occasionally, determination of properties of the aqueous solution in equilibrium with the solid, such as pH, conductivity, or concentration of ionic species is also of interest—in particular, in the monitoring of cleaning and consolidation of archaeological artifacts.

Chemical and morphological information can be combined by scanning lines or areas of the sample with the appropriate detection system. Compositional mapping of the studied region of the sample is then obtained.

Aging studies, performed in the laboratory, are useful for confirming theoretical models describing the behavior of the object at short-, medium-, and long-term intervals. Formed alteration products, (e.g., by oxidation, reduction, polymerization, scission, hydration, dehydration, dehydrogenation, etc.) are the target compounds in such studies. Three-dimensional (3D) diagrams can be built from the spectra or other characteristic curves obtained at different times.

(d) Biological information. Biological studies focus on the identification of the attacking species (fungi, algae, or bacterial microorganisms, as well as insects, plants, etc. ...), the products resulting from their metabolic activity, and the alteration products resulting from their interaction with the object. Identification of microorganisms and quantification of species are the tasks of primary studies. Following that, reaction pathways, as result of metabolic activity, are established from data provided through instrumental techniques. Morphological and chemical changes of the object due to biological activity are determined according to that described earlier.

1.2.2 Analytical Information Obtained from the Environment

The term "environment" includes both general and local conditions surrounding the object. According to that definition, three main types of environment can be established: aerial, terrestrial, and underwater. Most of the cultural goods are exposed to the atmosphere, and therefore, determination of physical and chemical parameter characteristics of this environment is of interest when attempting to accurately establish the causes of the damage exhibited on the object. In outdoor conditions, water in the atmosphere (precipitation, humidity, condensation) is an important data for characterizing the effect of the environment on the object. Determination of the content of pollutants in the air (CO_2, NO_x, SO_2, O_3, organic compounds, marine aerosols, suspended matter, etc.), along with the determination of temperature, are also commonly required for characterizing the aerial environment to which the object is exposed. In indoor conditions, emanations from furniture and exhibition cases are investigated, together with most of the above-mentioned parameters. Other environmental factors to be considered are solarization, microorganisms, wind and rain regime, and vibrations caused by road, rail, and traffic.

A large number of archaeological artifacts are found in burial conditions. Water acting as a solvent, as well as a carrier of ionic species coming from the soil, is responsible for the migration of the latter—the acid/alkaline attack on the object material and further lixiviation of materials and ionic species from the object. Thus, determination of physical and chemical properties of the soil (pH, conductivity, chemical composition, etc.) is of great importance.

Finally, underwater and waterlogged environments require a complete chemical analysis of the water (pH, content of ionic species, etc.), as well as the determination of its physical properties (temperature, density, conductivity).

1.2.3 Analytical Information Obtained from the Conservation Process

Analytical control of restoration, conservation, and preservation processes is increasingly demanded by personnel in charge of these protective tasks who attempt to carry them out in the most appropriate mode. Analytical information involving these activities varies widely depending on the methodology employed in the intervention, the chemical products used, and the class of material treated. Cleaning operations on ceramics and (less frequently) on stone or plasters require successive measurement of conductivity in the cleaning bath until the content of soluble salts in the object is reduced to a specific low value. Loss of material from the object and, in general, changes in morphology and chemical composition are controlled by a comparison of the analytical data obtained during the intervention by using a number of microscopic, spectroscopic, spectrometric, chromatographic and electrochemical techniques, among others. Changes in the visual appearance of the object (in particular, color) are controlled by spectrophotometric data obtained from the surface of the object. Sometimes lixiviated and extracted materials are determined during cleaning. In parallel, the chemical and physical properties of the products used for cleaning are controlled. Mechanical cleaning, or cleaning by means of laser ablation systems, can also be analytically controlled.

Similar to cleaning operations, consolidation and adhesion treatments are usually under analytical control. Measurements of the chemical, morphological, and physical properties of the consolidated material are indispensable to assure the efficiency of the operation. Changes in mechanical strength, color, and gloss, as well as chemical composition are assessed. Studies of stability of consolidants are frequently performed by means of natural and accelerated aging trials.

Application of finishes or protective coatings and products for casting and filling missing parts is also surveyed analytically. Compatibility among the products applied and the original materials is assessed from a chemical and physical point of view. Likewise with the consolidants and adhesives, accelerated aging trials are applied to check the stability of the products proposed as coatings and fillers. Experimental conditions for their application—in particular, temperature and humidity—are considered as they determine the final results of the treatment.

Resistance to biodeterioration is another factor to be considered when a new material is proposed for conservation purposes. Accelerated aging trials on inoculated specimens are developed in an attempt to characterize the behavior of this product from the point of view of its resistance to the biological attacks.

Physical conditions of transport and storage of works of art require analytical controls in order to guarantee the preservation of the object. Temperature, humidity, and vibrations, among other parameters, are considered in such instances.

1.3 Requirements of Analytical Methodology Applied to Archaeometric and Conservation Research

The premise establishing that each cultural good is unique and irreplaceable, and as consequence must be preserved as intact as possible, restricts the conditions for applying analytical procedures. Thus, the four steps comprised in the complete process of analysis, as shown in Scheme 1.3, are conditioned by this singular character inherent to monuments, art, and archaeological objects.

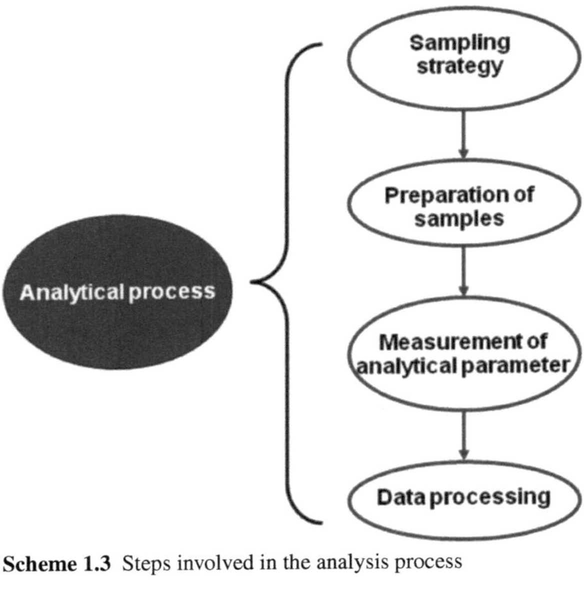

Scheme 1.3 Steps involved in the analysis process

1.3.1 Sampling Strategy

Sampling strategy is the first step in the analysis process, in which the analyst must make a number of decisions about location of sampling points, method of sampling, and the number and size of the samples. Concerning the former and, according to

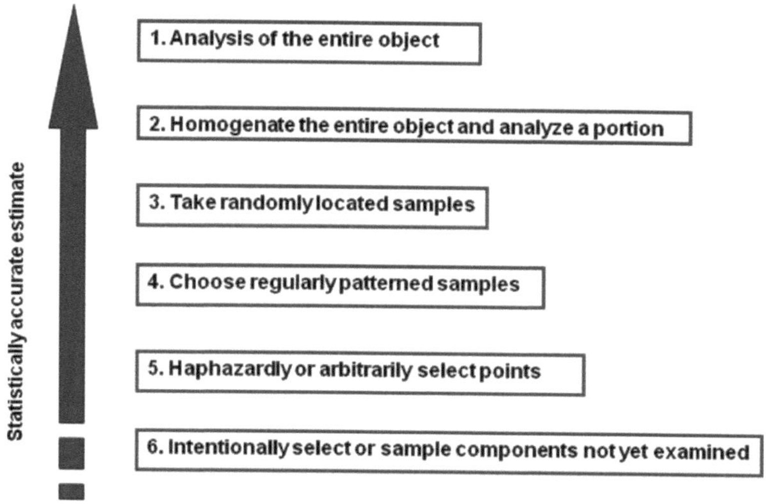

Scheme 1.4 Sampling strategies for the analysis of a single object

Reedy and Reedy [1], there are six possible sampling strategies for the analysis of a single art or archaeological object, as summarized in Scheme 1.4:

1. *Analysis of the entire object.* Sampling the entire object is usually impossible in archaeometric and art conservation studies, despite the fact that this is the best method for obtaining an accurate result. Nevertheless, some instrumental techniques that do not require sampling, such as photography or x-ray radiography of the entire artifact, are good examples of application of this strategy.
2. *Homogenate the entire object and analyze a portion.* Similar to the prior strategy, this method of sampling is usually impossible in archaeometric and art conservation studies.
3. *Take randomly located samples.* This strategy is equivalent to Strategy 2, since it provides an estimate of the composition of the entire object from a portion of the object. In contrast to Strategy 2, this method of sampling is applicable to cultural goods.
4. *Choose regularly patterned samples.* This strategy consists of taking samples at regular intervals across an object. The accuracy of the estimate is satisfactory for this strategy. Nevertheless, the risk of obtaining a biased result is comported in this strategy when a spatial pattern is present in the object at the same scale as the sampling interval.
5. *Haphazardly or arbitrarily select points.* This strategy imposes restrictions in sampling—in particular, positions for aesthetic or preservation reasons. Thus, there is the risk of obtaining a biased estimate when this strategy is applied on an object.
6. *Intentionally select or sample components not yet examined.* This strategy is frequently applied in art conservation research aimed at the identification of specific alterations or the characterization of the artist's palette.

The number of samples to take depends upon the type of object analyzed and the level of accuracy required for the estimate. In general, the larger the number of samples, the higher the level of accuracy is reached in the analysis; however, the number of samples is restricted by the constraints within the field of art conservation research, so a compromise between these requisites must be achieved.

The size of sample is also restricted in the field of art conservation research. Nevertheless, the degree of heterogeneity of the object imposes an inferior limit to the sample's size, so a compromise between size of sample and size of the confidence interval acceptable should be achieved in each particular case study.

Finally, the sampling method is also conditioned by the singular character of cultural goods. Analytical techniques not requiring sampling are preferred in art conservation research, but they do not always provide the necessary information on the composition of the object so therefore sampling must be carried out. The method of sampling depends on the analytical technique chosen for performing the analysis. Mechanical excision of samples in micro to nano scale is frequently carried out. Restrictions in the size of samples impose the use of scalpels, lancets, needles and, less frequently, instruments specially designed for this purposes. Sampling is, sometimes, performed by abrasively transferring a few grains of the solid material from the object onto a small SiC disk, which does not interfere when techniques such as attenuated total reflectance Fourier transform infrared spectroscopy (ATR-FTIR) are used for analysis. Samples of soluble materials can be taken by dissolving them with an appropriate solvent. In such instances, a cottons swab is impregnated with solvent, and then is rolled on the surface of the object until complete extraction of the soluble materials (which are retained in the swab) is complete. The materials retained in the swab are redissolved and then the solution obtained is analyzed. Less frequently, they are directly analyzed in the swab (i.e., by means of pyrolysis gas chromatography mass spectrometry, Py-GC-MS).

1.3.2 Preparation of Samples

Most of the analytical techniques applied in art conservation research require the preparation of the sample prior to the analysis step. Although the sample preparation procedures vary in a wide range, five basic types of procedures can be established: grinding, dissolving, derivatizing, melting and embedding.

Powdering, or grinding, of samples is a simple preparation method required in a number of spectrometric and spectroscopic techniques, such as x-ray diffraction (XRD), nuclear magnetic resonance (NMR), differential thermal analysis (DTA), thermogravimetric analysis (TG), or ATR-FTIR spectroscopy. Control of the particle size during grinding must be taken into account in attempting to obtain reliable results.

Dissolution of the sample is the method required in a number of spectroscopic and chromatographic techniques (e.g., UV-Vis spectrophotometry, atomic absorption spectroscopy (AAS), high performance liquid chromatography (HPLC), and thin-layer chromatography (TLC)). Selection of the suitable solvent is essential when this preparation methodology is used. Occasionally, it is necessary to perform an acid or alkaline attack for dissolving the sample.

Application of gas chromatographic techniques is restricted by the necessary volatility that the analyzed compounds should exhibit. A derivatization reagent is added, in a prior step, to render volatile the components of the analyzed organic materials. In some cases, a prior step consisting of an acid or alkaline hydrolysis of the organic material is necessary for releasing the molecular constituents of the polymeric structure. Inclusion of a prior step in the preparation procedure devoted to the suppression of interfering species is sometimes included.

Melting of samples is necessary for performing the analysis of ceramics and glass materials by means of x-ray fluorescence (XRF). Lithium tetraborate is added as flux for lowering the melting temperature. The homogeneous disks that form can be considered a solid solution of the sample compounds in the binder.

Finally, microscopic examination of samples often requires their preparation as cross sections or thin sections, or by mounting the sample on a glass slide by means of a mounting medium. For preparing thin and cross sections, samples are embedded in a polymer solution. After curing of the polymer, the thin or cross section is obtained by polishing the embedded sample with SiC abrasive disks. Aluminum suspensions or diamond paste are occasionally employed in a final polishing step.

1.3.3 Measurement of Analytical Parameters

At the beginning of the analytical process the analyst has to select the method of analysis. That at least partly determines the sampling strategy, and it completely determines the preparation of the sample.

Two basic requirements must be met for the instrumental technique when it is applied in art conservation research: sensitivity, for obtaining relevant data from small samples on the nano, micro or mili (-gram, -meter) scale; and specificity, for unambiguously identifying compounds and quantifying the analytes from the complex mixtures of substances that form the materials present in the monument or artwork. Other requirements are also desirable for an analytical method when it is applied to objects of artistic, historic, and archaeological nature: according to Lahanier et al. [2], these are:

- nonintrusiveness of the analytical method so that physical integrity of the object is safeguarded
- nondestructiveness of the analytical method so that the bulk of the sample is recoverable after analysis
- fastness, allowing the analysis of single objects as well as large assemblages of them
- universality, enabling the analysis of materials and objects of various shapes, sizes, and compositions with minimum sampling amounts and a minimum of pretreatment

- versatility, to provide information on both average or bulk composition of the object and the specific composition of local areas
- multielement analysis capability, allowing qualitative information on multiple elements or compounds present in the object by means of a single measurement

1.3.4 Data Processing

Chemometric techniques are, nowadays, a valuable tool for achieving a more accurate characterization of the materials composing the art object. Whereas descriptive statistics are commonly used in conservation research, estimation methods are less frequently used. Among them, linear regression methods are mainly used in specific case studies. Multivariate techniques such as cluster analysis and discriminant analysis have been increasingly applied to case studies in archaeometric and conservation research over the last few decades. T-test and analysis of variance are the two hypothesis-testing methods most commonly used in art conservation research. An increasing number of papers have appeared in specialized literature in which fuzzy logic and data-fusion techniques have been applied to archaeometric studies [3].

1.4 An Overview on Analytical Methods Applied in Archaeometry and Conservation of Cultural Goods

As a result of the progressive practical application of the ideas of art historians like Johann Wincklemann (1717–1768) on art and technical history, a scientific approach to art and archaeological objects commenced around 1780 in parallel to that based on a text-based methodology [4]. At the beginning of the 20th century, analytical techniques were restricted to spot tests, optical microscopy, and some spectroscopy techniques. Nevertheless, the number of analytical techniques has significantly increased in the last few decades, extending the scope of classical chemical and physical analysis. Nowadays, there are a wide variety of scientific methodologies providing information on morphology, chemical composition, and structure of the materials present in the monument, archaeological artifact, or art object.

Application of scientific methods to archaeometry and the conservation of cultural heritage is carefully carried out to ensure that the methods chosen are in line with the purposes of the research. According to Lahanier [5], methods currently available are classified into three categories:

- Examinations based on recording images seen by electromagnetic radiations or electrons
- Analytical methods
- Dating methods

1.4.1 *Examination Based on Recording Images: The Holistic Approach*

This approach includes images of the object as a whole, visible to the naked eye, acquired in the range of visible, ultraviolet, and infrared light, x-rays, beta rays, and gamma rays, as summarized in Scheme 1.5. These methods work well with the manner in which art historians or conservators approach their work. As there is no sample extraction requirement, these methods can be considered nondestructive. Although, in a strict sense, they can be destructive—i.e., radiation could damage an object. In general, these technologies have been applied to the study of materials of archaeological, cultural, and historic value, monitoring and evaluation of conservation treatments, and digital imaging for documentation and archiving.

These techniques are usually classified according to the type of radiation or spectral region in which data are provided, namely, electromagnetic radiation (x-ray, UV, visible, IR, radio, etc.), acoustic radiation, etc.

(a) Electromagnetic radiation in the visible region. Digital cameras and high-resolution digital scans have progressively replaced more conventional photographic equipment as new documentation and recording techniques. Photogrammetric techniques are based on obtaining orthophotographic images and clouds of 2D or 3D scanned points (matrix-oriented or scattered) by means of digital cameras or laser scanners, which use different digitizing strategies. While photogrammetry and metric surveying techniques can be suitable for archaeological sites and buildings, they present certain disadvantages for smaller and more complex objects. These techniques, together with the development of image processing and image analysis approaches, combined with classification strategies based on fuzzy logic, have extended the scope of application of photographic techniques to a variety of fields, such as authentication of artwork or 3D virtual restoration [6].

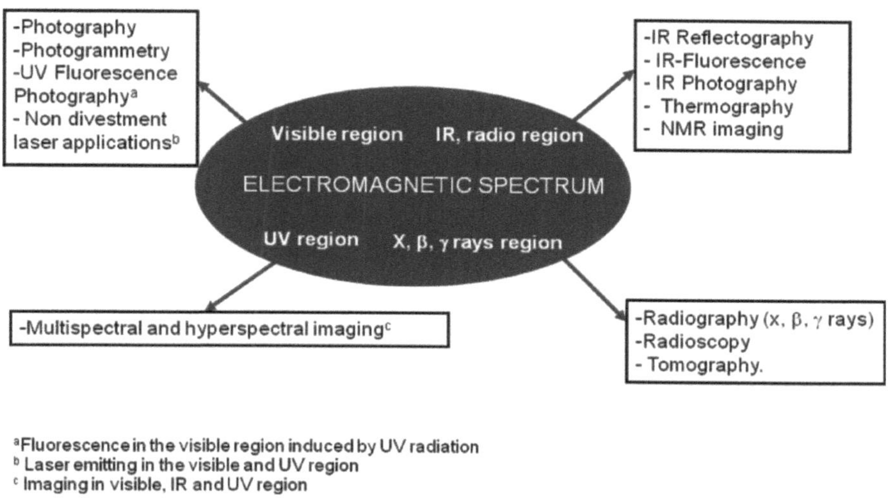

Scheme 1.5 Main analytical methods based on recording images seen by electromagnetic radiation

Ultraviolet light and fluorescence is a common diagnostic tool for the examination of painted surfaces and for monitoring cleaning processes used since 1925 when UV lamps were commercially available. This method provides information on the composition and specific characteristics of the paint surface, such as retouching and overpainting. Fluorescence lifetime imaging (FLIM) provides images of the fluorescence induced by two lasers emitting in the visible and UV ranges and recovered by a nanosecond time-gated intensified charge coupled device (CCD) camera [7].

Other real and potential uses for lasers in art conservation analyses and practice have been investigated over the last few decades. They are so-called nondivestment laser applications, and are being increasingly used in artwork conservation. [8] These include holographic interferometry nondestructive testing (HINDT), speckle pattern interferometry (SPI), and speckle pattern shearography (SPS), also called grazing-incidence laser scattering.

(b) Electromagnetic radiation in other spectral regions. Infrared photography has been routinely used to examine paintings with films sensitive in the spectral range of 700–900 nm since the 1930s. The obtained images provide interesting information on the working procedures and presence of retouching and overpainting, and enable the identification of underdrawing and the detection of changes in the composition of the materials used by the artist. In 1960s reflectography, using radiation in the wave-length range of 1000–2000 nm was introduced, providing greater penetration through the object strata, and thus enabling a better study of underdrawings. Reflectographic equipment is currently coupled to TV monitors, digitized video cameras, solid-state cameras based on either PtSi and InGaAS sensors, and high-resolution infrared cameras [9].

Computer controllable spectral-imaging techniques measure optical reflectance spectra at many points on a target simultaneously, producing a digital stack of information on defined areas of an object exposed to many different wavelengths of light. Originally described as "multispectral" imaging, the continuous increase in the number of spectral bands resolved by the new generations of imaging spectrometers resulted in the appearance of the term "hyperspectral." The system records light intensity as a function of both wavelength and location, so that a full image at each individual wavelength is included in the data set and thus extends the capabilities for diagnostic imaging [10].

X-rays discovered by Röntgen in 1895 are the foundation of noninvasive radiological techniques. The image formed by the x-ray emitted from a x-ray source, and transmitted through an object, is observed on a fluorescent screen using radioscopic techniques, registered on radiographic films using radiographic techniques (Fig. 1.4), or registered as digital images. The voltage applied in the x-ray tube depends on the characteristics of the studied object. Gammagraphy techniques use radioactive isotopes as a source of more energetic gamma rays [11].

Computed X-ray tomography (CT) was developed as a radiological imaging technique in the 1970s. Three-dimensional CT images in any plane can be reconstructed from a set of sequential cross-sectional slices with resolution of 0.1 mm. Main applications of this technique involve monitoring of the deterioration of natural building stones. More recently, x-ray computed microtomographs (μCT) achieve higher resolution so that mineral grains, micropores, or cracks in the range

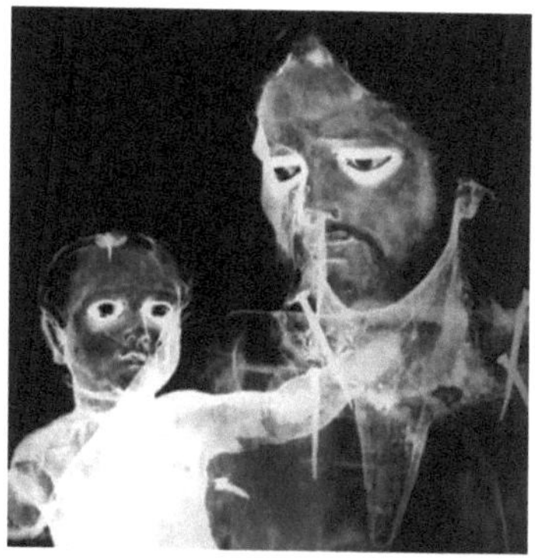

Fig. 1.4 Radiograph of a polychromed sculpture. Internal nails used for assembling the different parts of the sculpture are clearly viewed. Courtesy of J.A. Madrid, Instituto de Restauración del Patrimonio, Universidad Politécnica de Valencia, Spain

of 10–100 μm can be detected. This technique has been successfully applied to the examination of paintings and to the characterization of buried archaeological glasses [12].

Magnetic resonance imaging (MRI) has been applied to the study of the distribution of fluid components (i.e., water or a polymer used as consolidant) in a porous material such as stone or waterlogged wood by a direct visualization of the water or fluid confined in the opaque porous medium [13].

(c) Other radiations. Acoustic emission (AE) based on the energy released due to microdisplacements in a structure undergoing deformation, which is detected on the surface using a piezoelectric transducer, has been used for studying the alterations and damage on wooden objects [14].

Laser-induced ultrasonic imaging is used quite widely to study the interiors of opaque objects such as sculptures and paintings from the exploitation of laser-induced stress waves [8].

In laser Doppler vibrometry (SLDV), surfaces are slightly vibrated by mechanical activation while the vibrometer scans the object producing 2- or 3D maps of velocity, amplitude, and phase, which allow the detection and mapping of structural defects [15].

Thermography using heat radiation enables the observation of temperature differences, thermal emissivity, or thermal conduction of the studied object. This technique is mainly used in the examination of monuments [16].

1.4.2 Analytical Methods: The Reductionist Approach

Characterization of the objects based on the holistic approach is often combined with point analysis, in which the material composition of a minute portion of the object is determined by classical analytical procedures or instrumental methods. Since the samples under examination are unique and irreplaceable, the specimens should be submitted to nondestructive or at least microdestructive analysis. Nevertheless, a number of instrumental techniques require the total or partial destruction of the

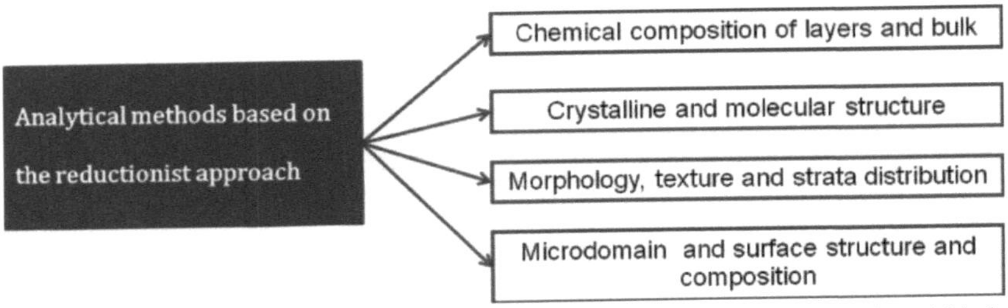

Scheme 1.6 Classification of point analysis methods according to the information yielded

sample. Thus, depending on the procedure involved, the analysis can be considered destructive or nondestructive, and is carried out on the bulk or the object surface. Moreover, and depending on the instrumental technique used, the obtained data can be panoramic or sequential and the measurements can be directly performed on the work itself or on a sample.

According to Mairinger and Schreiner [17], chemical and physical methods can be classified in four categories, as illustrated in Scheme 1.6.

1.4.3 Point analysis providing chemical composition of layers and bulk

Chemical methods and instrumental techniques included in this group provide elemental composition and characterization of ionic species present in the bulk or layers of the object. According to the physical or chemical principia in which the method is based, it can be classified as illustrated in Scheme 1.7:

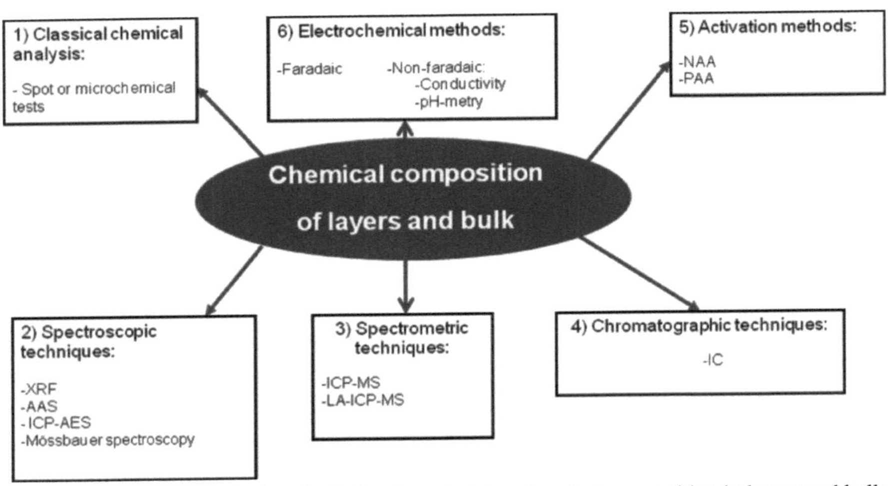

Scheme 1.7 Main analytical methods for characterizing chemical composition in layers and bulk

(a) Classical chemical analysis The first chemical investigations on the components of historical materials, carried out in the late 18th century, were based on the classical analytical procedures. At that early time, microchemical tests were applied to identify pigments and inorganic materials. During the 20th century, microchemical tests based on classical analytical procedures have been frequently used in specialized laboratories. They not only can serve to calibrate instrumental methods but also they provide a quick and sensitive method of identifying ionic species in materials. In addition, they have, as another advantage, the feasibility of identifying the efflorescence crusts or the pigment within the paint layer, as these tests can be directly performed on cross sections, and the reaction can be observed with the help of a stereomicroscope [18].

(b) Spectroscopic techniques. Spectroscopic techniques have been widely used in the identification and determination of artist's materials of both organic and inorganic types [19]. The provided information and the application of each specific technique depends on the range of electromagnetic radiation and the phenomenon involved in its interaction with the materials present in the analyzed object.

X-ray spectrometric techniques involve radiation in the wavelength range from about 10^{-4} to about 10 nm. These techniques are based on the identification of the characteristic x-rays emitted by the atoms of a sample irradiated by a sufficiently high energy source. Portable instruments enabling in situ investigations, as well as alternative techniques such as micro-XRF, polychromatic synchrotron micro-XRF, total reflection XRF or x-ray absorption have been proposed in an attempt to improve the obtained results [20]. The main advantages of XRF techniques are their rapidily enabling multicomponent analysis, and providing simple spectra, accuracy, and reproducibility. Applications of this technique to the analysis of artwork have been related to the identification and determination of major, minor, and trace elements composing inorganic materials such as pigments, metal, stone, ceramics, glass, surface coatings, and deposits of adventitious materials on the surface, etc. [21].

Emission spectroscopic techniques, using spark source or continuous direct current (DC) arc to excite the emission lines of the elements, and inductively coupled plasma-atomic emission spectroscopy (ICP-AES) together with AAS, have been widely applied in the detection and determination of the chemical composition of inorganic materials. They are especially good at detecting trace elements so they were employed for many of the major analytical studies of glass, glazes, and alloys among other antiquities and archaeological materials in earlier decades [22]. In contrast to other spectroscopic techniques, these are intrinsically destructive and require specific analytical procedures and calibration standards for each analyte.

Mössbauer spectroscopy is a selective tool for the quantitative analysis and speciation of a very limited number of elements. It has been mainly used to study iron compounds—e.g., ceramics, as it gives valuable information about iron-bearing oxide and silicate minerals. This technique has been applied to the identification of the provenance of clay and used raw materials—the manufacturing method employed in pottery and, to a lesser extent, to the characterization of pigments and weathering crusts formed on stone monuments [23].

(c) Spectrometric techniques. Inductively coupled plasma-mass spectrometry (ICP-MS) is a relatively new analytical technique developed during the 1980s and increasingly used in earth sciences and archaeometry fields. Its recent combination with laser ablation to create inductively coupled plasma-mass spectrometry (LA-ICPMS) has increased the capabilities of this technique. A large number of analyzed elements (10–30 typically), a minimal amount of material (tens of mg or less), and the rather simple pretreatment required, as well as the feasibility of performing isotope ratio measurements are the main advantages of this technique. This technique has been applied to the trace element characterization and source tracing of obsidian (*vide infra*), compositional analysis of metallic artifacts, and lead isotope ratio analysis [24].

(d) Chromatographic techniques. Ion chromatography (IC) is a convenient technique for identifying and determining the content of soluble salts on metals, stone and ceramics materials used in monuments—in particular, the anions—and to monitor the ionic species removed during water immersion treatments of archaeological ceramic remains [25].

(e) Activation methods. Activation analysis (AA) has become a popular analytical technique in archaeology, which has provided interesting data enabling the elucidation of the major, minor, and trace element composition of art and archaeological artifacts, which, in turn, can be used to establish their provenance and temporal origin. Detection limits and the range of elements analyzed depends on the type of activation—namely, neutron, fast neutron, charged particle, or gamma photon.

Neutron activation analysis (NAA), based on the interaction of the object material with fast neutrons, has been used in the identification and determination of the content of elements present in pigments, coins and alloys, stone, glass, and pottery [26]. Multi-elemental analysis (about 20 elements) can be performed on small samples off less than 5 mg, with sensitivities in the ppm range. The requirement of a nuclear reactor, the handling of radioactive materials, and the time-consuming procedures required for preparing the samples are the main drawbacks of this technique.

Proton activation analysis (PAA) provides chemical composition of the materials at a depth of 300–500 μm under the irradiated surface. This technique has been used in the study of metallic objects such as ancient coins [27].

(f) Electrochemical methods.

(I) Faradaic electrochemical methods. From a general analytical point of view, electrochemical techniques are very sensitive methods for identifying and determining the electroactive species present in the sample and, in addition, they also are able to carry out speciation studies, providing a complete description of the states of oxidation in which the ionic species are present in the object. Other applications and improvements obtained by their hyphenation with other instrumental techniques, such as atomic force microscopy (AFM), will be described in the following chapters.

(II) Non-Faradaic electrochemical methods. Conductometric methods have been extensively used by scientists and conservators for monitoring the content of salts removed during water immersion treatments of ancient tiles and archaeological ceramic remains. In a different manner to IC, this technique provides the total ionic

content of ionic species present in the cleaning bath. Its main advantage is the simplicity of the instruments [28].

Measurement of pH is a potentiometric technique frequently used for measuring the degree of the deterioration of materials that are subjected to natural aging. The determination of pH levels is commonly carried out on ethnographic objects manufactured with parchment or leather, and it is especially relevant in altered paper due to the formation of acidic compounds from the decomposition of the woodpulps and other raw materials, which can induce the hydrolysis of the cellulose and then decrease the resistance and mechanical properties of the document [29].

1.4.4 Point analysis providing molecular and crystalline structure

An overview of the analytical techniques most frequently used that provide molecular and crystalline structure is illustrated in Scheme 1.8. Basically, they can be grouped into histochemical and immunological methods, diffraction, spectroscopic, spectrometric, chromatographic, and thermoanalytical techniques.

(a) Histochemical and immunological analysis. As for spot tests applied to pigments, the first investigations of the organic components of historical materials were carried out in the late 18th century and consisted of solubility tests, tests for determining some characteristic physical property of the medium (such as the melting point), and, preferably, histochemical tests [4]. Application of histochemical tests to binding media identification is limited, however, because of its low specificity (i.e., discrimination between albuminoids, casein, or gelatin within proteinaceous materials is hardly obtained). Despite that, a noticeable number of methods have been proposed in the literature [30].

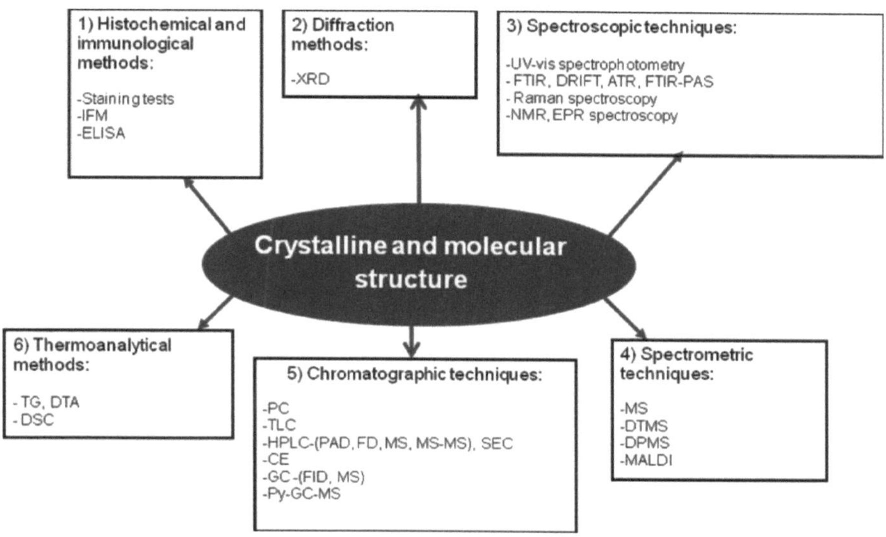

Scheme 1.8 Main analytical methods for characterizing crystalline and molecular structure

Further improvement of microchemical methods for proteinaceous media was based on immunological techniques. The high specificity of the antigen-antibody reaction enables the discrimination of the same protein coming from different species, or the detection of multiple antigens in the same sample. Application to the analysis of artwork has been reported in two types of immunological techniques: immunofluorescence microscopy (IFM), and enzyme-linked immunosorbent assays (ELISA) [31].

(b) Diffraction techniques. X-ray diffraction has been an excellent tool in the service of art and art conservation, as it provides essential information on mineralogical composition and the crystalline structure of materials. Thus, a number of monographs can be found devoted to the description of historic pigments, which include data concerning the x-ray diffraction patterns of many minerals and pigments from historic collections [32]. In addition to pigments, a wide variety of materials can be identified by means of this instrumental technique: stone, clays and ceramic bodies, glazes, efflorescences, and corrosion products formed on metallic objects.

New portable spectrometers have been used for in situ analyses. Powder x-ray microdiffraction instruments or those with highly collimated and parallel x-ray beams have also been used for characterizing cultural goods.

(c) Spectroscopic techniques. Application of the UV-Vis spectrophotometry solution to the study of dyes dates back to 1889 and has been further used in the identification of these substances [33]. Use of this technique for analyzing organic binders is notably restricted because of the nonspecific spectra provided by these materials. In parallel, instrumentation providing visible region reflectance spectra and color specification using colorimetry according to the Comission Internacional de l'Eclairage (CIE) and hand-held spectrophotometers has been a common tool in the field of art and art conservation for in situ characterization of materials such as wall painting pigments or glazed ceramics [34].

Photoluminescence spectroscopy provides information on the composition of the painting surface and the presence of retouchings and overpainting. This technique has progressively evolved using laser sources, giving rise to laser-induced fluorescence (LIF) spectroscopy [35].

Infrared spectroscopy is a widely used technique in the field of art and art conservation due to its versatility and ability to provide structural information on both inorganic and organic materials, the small amount of sample required, and minimum pretreatment.

Since its introduction in the field of conservation science in the 1950s, an increasing number of studies have been reported in the technical literature, as illustrated in some reviews [19, 36]. The noticeable improvements in the instrumentation—in particular, the incorporation of IR Fourier transform spectroscopy—have significantly contributed to extending the scope of this technique so that nowadays it is routinely used for analyzing art and archaeological objects, monitoring conservation treatments, and assessing new conservation products and methods. A wide range of FTIR methods are currently in use: diffuse reflection Fourier-transform infrared spectroscopy (DRIFT), ATR (Fig. 1.5), FTIR photoacoustic spectroscopy (FTIR-PAS), and FTIR microspectroscopy (*vide infra*), either in the transmission or reflection mode [37].

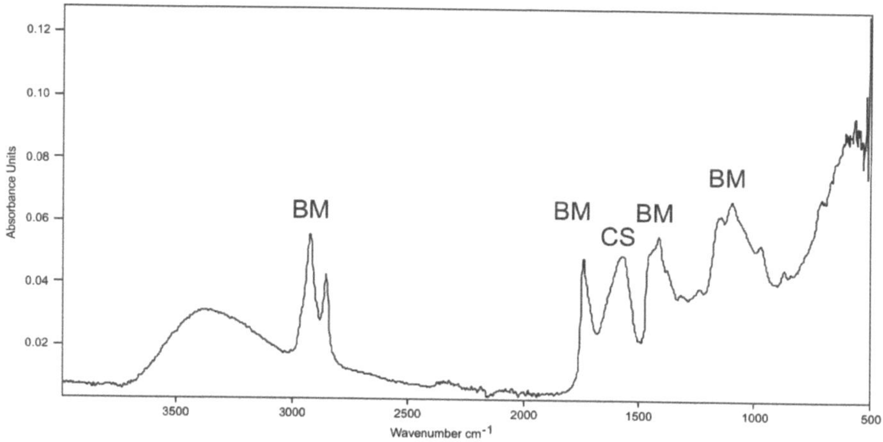

Fig. 1.5 IR spectrum, obtained in ATR mode, of a sample of yellow pigment from José Benlliure's palette (1937). (BM) IR bands ascribed to the drying oil used as a binding medium. Interestingly, and (CS) appearing at 1569 cm^{-1} is associated with carboxylate groups from cadmium soaps formed as a consequence of natural aging

Similarly to FTIR spectroscopy, Raman spectroscopy is a versatile technique of analyzing both organic and inorganic materials that has experienced noticeable growth in the field of art and art conservation, in parallel to the improvement of the instrumentation [38]. In particular, the introduction of fiber optic devices has made feasible the development of mobile Raman equipments, enabling nondestructive in situ analyses [39]. On the other hand, the coupling of Raman spectroscopes with optical microscopes has given rise to Raman microscopy (*vide infra*).

NMR has been primarily applied in archaeometric studies [40]. In contrast, NMR has had a restricted application in the art conservation field due to the complexity of the paint samples. This technique has been chiefly used for identifying highly polymeric materials, such as triterpenoid varnishes, oil, oleoresinous media, and synthetic media [41].

Electron paramagnetic resonance spectroscopy (EPR) (also called electron spin resonance spectroscopy, ESR) has been scarcely applied in the field of art and art conservation. Some work can be found in which EPR is used as complementary technique to SEM-EDX, NMR, and mass spectrometry (MS) for studying free radicals occurring in polymerization, pyrolytic, oxidative, and other radical degradative processes in artwork, as well as in the characterization of varnishes and oleoresinous media [42].

(d) Spectrometric techniques. Mass spectrometry (MS) is an excellent tool at the service of the art and art conservation field due to its great potential to resolve molecular structures without sample pretreatments. Nevertheless, application of this technique in conservation practice is limited by the complexity of art materials. This drawback can be resolved by hyphenation of MS with a chromatographic method. Despite this, a noticeable number of works can be found in literature using, preferably, three different MS techniques. Direct infusion MS has been applied in the analysis of proteinaceous and oil media [43]. Direct pyrolysis MS (DPMS) has been

applied to the identification of natural gums, resins, and waxes in ancient Egyptian mummy cases, as well as in the study of thermally degraded products formed from traditional binding media [44]. Direct temperature-resolved MS (DTMS) has been applied to the analysis of traditional binding media and synthetic varnishes [45].

The introduction of matrices to assist the laser desorption ionization (LDI) process has lead to the development of the so-called matrix-assisted LDI (MALDI). This technique has been applied to identify terpenoid varnishes and their oxidized products [46]. Combined with enzymatic cleavage, MALDI has also been used in the identification of animal glue.

(e) Chromatographic techniques. Chromatographic techniques are commonly used in the analysis of artwork due to their ability to separate the organic components of the complex mixtures present in objects of historic, artistic, and archaeological nature. In general, they not only achieve a higher discrimination of the type of material present in the object, but they also identify the alteration products of these materials. Technological advances have determined the evolution of these techniques in the field of art and art conservation. Thus, gas chromatography (GC) or liquid chromatography (LC) have progressively replaced other simpler techniques, such as paper chromatography (PC) or TLC. Hyphenation of GC and LC to a MS detector, as well as use of pyrolysis devices coupled to the GC, have significantly improved and extended the scope of applying these techniques. Complex and time-consuming sample preparation [47] is the main drawback of these techniques, sometimes limiting their application—in particular, when a large number of measurements must be periodically performed on large assemblages of specimens (i.e., laboratory experiments on monitoring alteration processes of artists' materials or conservation products). Despite that, these techniques are abundantly used in specialized laboratories for analyzing organic materials such as organic pigments, binding media, and coatings.

Reverse-phase columns with a gradient elution in combination with UV-Vis spectrophotometers using photodiode-array (PDA) (Fig. 1.6) and spectrofluorimeters are common devices employed in this technique. In a lesser extent, MS, tandem mass spectrometry (MS-MS), and nano liquid chromatography-electrospray ionization-quadrupole time-of-flight tandem mass spectrometry (nanoLC-nanoESI-Q-qTOF-MS-MS) has been used as detection system. This instrumentation has been mainly used in the analysis of dyes and proteinaceous media, and in some extent, in the analysis of drying oils and terpenoid varnishes [47, 48].

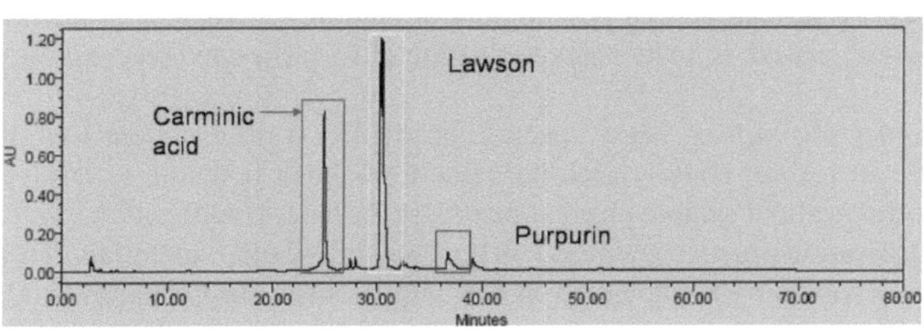

Fig. 1.6 Liquid chromatogram from a mixture of compounds responsible for the color of cochineal henna, and madder. Courtesy by DJ Yusá, Instituto de Restauración del Patrimonio, Universidad Politécnica de Valencia, Spain

Size-exclusion chromatography (SEC) has been also used as complementary technique for characterizing changes in the composition of synthetic resins due to aging. In particular, it enables the determination of molecular weight distributions and the glass transition temperatures of resins.

In recent years, capillary electrophoresis (CE) has been introduced as an alternative method for performing a chromatographic analysis of art objects, and thus, a number of methods for analysis of waxes, resins, drying oils, animal glues, and plant gums have been proposed [49].

GC has been extensively applied in the identification of organic media and varnishes given its versatility. In the first analytical studies of artworks, flame ionization detectors (FID) were used. The higher capacity of MS detectors—coupled with the GC for identifying compounds—and their high sensitivity are the reasons for the extended use of GC-MS instruments, preferably quadrupole mass spectrometers and ion-trap mass spectrometers. Specific derivatization methods have been proposed according to the type of target compound: polysacchraides, oils and waxes, terpenoid resins, or proteinaceous media [47]. Additionally, a number of papers can be found in the literature in which more sophisticated pretreatments are proposed to identify more than one class of compounds in the same paint sample [50].

The combination of pyrolysis, gas chromatography (Py-GC) and MS (Py-GC-MS) has notably improved the identification of organic artists' materials and, in particular, has extended the scope of GC to the identification of synthetic materials. Thus, this technique has become a powerful tool for scientists involved in art and art conservation. Py-GC-MS offers a number of advantages: minimum sample preparation—it can be directly analyzed in a solid or liquid state with no hydrolysis or derivatization pretreatment and high sensitivity and LOD (below the μg). In contrast, the obtained pyrograms are more complex than the chromatograms obtained with conventional GC-MS due to the new compounds appearing in the pyrogram as a result of the pyrolytic fragmentation. Recent progress in Py-GC-MS relies on the development of online derivatization methods. This technique has been mainly applied to the analysis of lipids, natural di- and triterpenoid resins, and synthetic resins [51].

(f) Thermoanalytical methods. TG and DTA have frequently been applied in the field of art and art conservation as complementary techniques together with chromatographic, spectroscopic, or spectrometric techniques. They provide interesting information concerning the crystalline structure of the target compound and the degree of hydration of inorganic salts. More occasionally, these techniques and differential scanning calorimetry (DSC) have been applied to the analysis of binding media because a number of works have developed methods for ascribing exothermal reactions to specific organic compounds [52].

1.4.5 Point Analysis Providing Morphology, Texture and Strata Distribution

Many of the instrumental methods yielding morphological, topological, and textural information of objects are based on microscopy techniques. Although light microscopy (LM) has pioneered the technical examination of art objects and antiquities, other microscopy techniques, such as electron microscopy or atomic force microscopy, have been further developed, which overcomes the magnification limitations of LM or the extremely shallow depth of field at high magnifications (Scheme 1.9).

Low-magnification LM is a valuable technique for the examination of art and archaeological objects, used to gather preliminary information on alterations and damages suffered by the object and its state of conservation. High-magnification LM is currently performed in the range of $\times 100$–$\times 1500$. The depth of focus is relatively small, requiring a time-consuming preparation of samples to obtain a suitable image of the specimen. Image analysis enables point counting, and stereological and densitometric studies. In particular, characteristics of materials such as a percentage of aggregates, pores, temper or specific minerals, pore or grain size, and grain shape can be determined, which allows a better analysis and interpretation of composition, technology, provenance, deterioration, and preservation [53]. Other optical microscopy techniques, such as confocal laser scanning microscopy (CLSM) have been scarcely applied to the stratigraphic study of cross sections and the examination of canvas fibers.

Electron microscopy is an efficient microscopy technique that has been extensively used for the material characterization of artistic and archaeological objects, especially in combination with x-ray microanalysis [54]. The use of electrons instead of light in these instruments is the basis of the higher resolution (~9–0.2 nm) and has greater depth of field than LM. Thus, characterization of the finest topography of the surface objects is possible, and additional analytical information can be obtained. Different electron microscopes are currently used in art and art conservation studies: scanning electron microscopes (SEM), Cryo-SEM

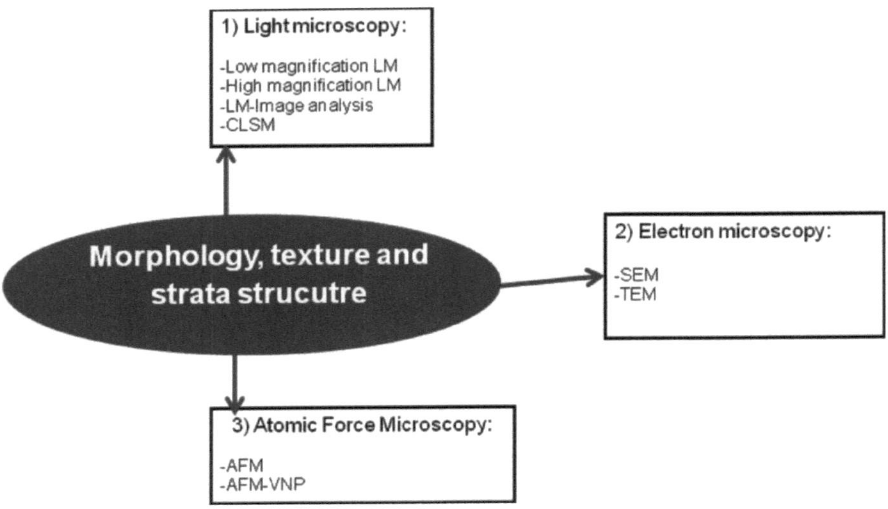

Scheme 1.9 Main analytical methods for morphological, textural, and strata distribution

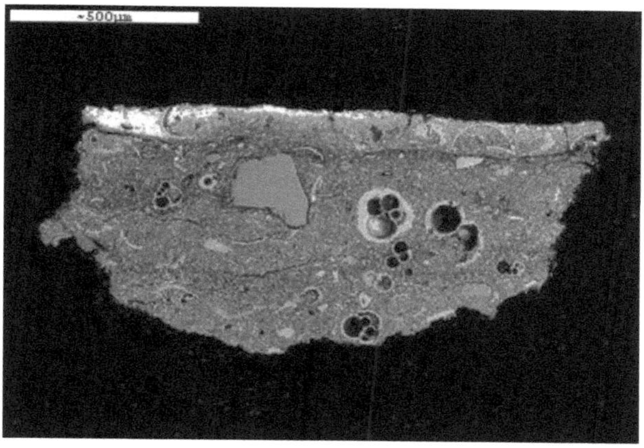

Fig. 1.7 Cross section of a painting (17th century, Taormina, Italy). Interestingly, the ground was prepared with a pigment obtained by crushing a biocalcarenite rock containing abundant microfossils (globigerine foraminifers, Sicily)

and environmental scanning electron microscopes (ESEM), and transmission electron microscope (TEM). These techniques have been widely used for studying the texture of materials (Fig. 1.7), the alteration and biodeterioration of materials, and for monitoring consolidation and cleaning treatments on stone, ceramics, paintings, sculpture, archaeological, and ethnographical objects, etc.

The AFM maps the topography of a substrate by monitoring the interaction force between the sample and a sharp tip attached to the end of a cantilever so that the morphology of the surface of the studied sample can be reproduced at nanometer resolution (Fig. 1.8). Some works can be found in literature, reporting studies in which AFM has been applied to the examination of art and archaeological objects [55].

1.4.6 Microbeam Analysis Providing Microdomain, Surface Structure, and Composition

In general, methods of surface investigation are based on the interaction of the incident energy provided by a microbeam of photons, electrons, or particles with the atoms or molecules located in the surface of the sample. Here the concept of "surface" should not be considered in a strict sense. Analytical interactions take place in the subjacent matter, achieving a depth in the range of a few micrometers. As result of the interaction between the incident beam and the sample, atoms or molecules release photons, electrons, ions, or neutral particles. These emitted corpuscles are carrying analytical information that is registered after they are conveniently detected. Moreover, the use of highly collimated microbeams for exciting the studied material results in a high spatial resolution, and therefore, low dimensions for the analyzed

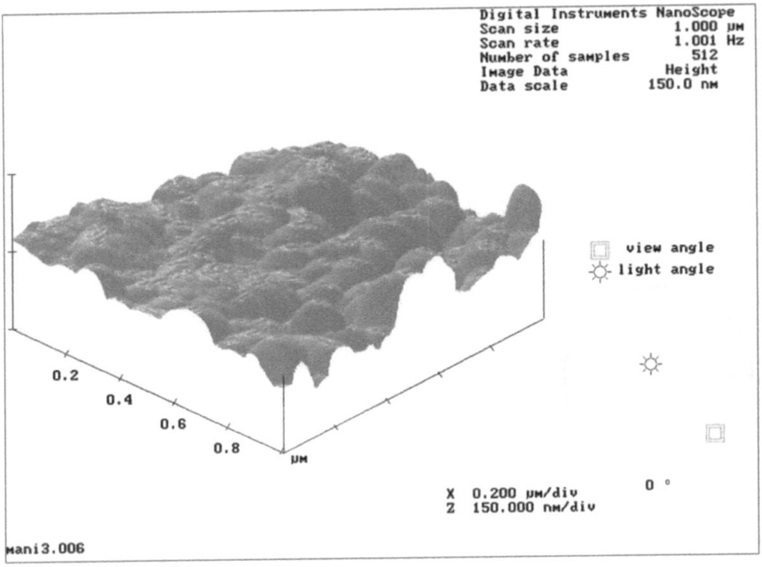

Fig. 1.8 Topographic map of the surface of an altered archaeological glass obtained by AFM

area. Table 1.1 summarizes the surface analysis techniques most frequently used in the characterization of art and archaeological objects.

Suitable high-resolution spatially resolved microspectroscopes operating in the vis region and IR region (FTIR and Raman) have been progressively applied to the analysis of art and archaeological samples providing spectral resolution for discriminating features in the different paint strata of a cross section [56–58]. In a lesser

Table 1.1 Classification of analysis surface techniques used in the study of cultural heritage according to the method of excitation and the resulting emitted corpuscles carrying analytical information

Excitation emission	Photon	Electron	Ion	Neutral particle
Photon	Vis light imaging microspectroscopy, microspectrofluorimetry, LIBS, Raman and FTIR imaging microspectroscopy, µXRF, µSRXRF, µXANES, XAS µSRXRD	SEM-EDX, EPMA		
Electron	XPS	AES, EELS		
Ion	PIXE, PIGE, LDMS, MALDI		RBS, SIMS	FAB

extent, microspectrofluorimetry and laser-induced breakdown spectroscopy (LIBS) have been applied to identify the type of binding media and the layer-by-layer pigment distribution, respectively [59].

Fast-atom bombardment mass spectrometry (FAB-MS) has been applied to the identification of diterpenoid compounds and their oxidation products. Similarly, laser-induced desorption mass spectrometric (LDMS) techniques have been applied to the identification of natural and synthetic organic pigments in microscopic paint samples prepared as cross sections [60].

Techniques of microscopic XRF (μ-XRF) developed in the last 20 years provide 2D images and elemental maps of each element present in the target material. Portable/in situ μ-XRF, μ-XRF spectrometers synchrotron-based (μ-SRXRF) and micro-x-ray absorption spectroscopy/micro-x-ray absorption near-edge structure spectroscopy (XAS/μ-XANES) have improved the mineralogical characterization, as well as the elemental and chemical imaging of samples at the submicrometer scale [61].

Synchroton-based micro-x-ray diffraction (μ-SRXRD) is being increasingly applied in the mineralogical characterization of art and archaeological objects [62].

External beam proton-induced x-ray emission (PIXE), and particle-induced gamma-ray emission (PIGE), are mainly applied to the characterization of porcelain, glass, glazes, pigments, gemstones, obsidian, bone and teeth, inks, stamps, coins, jewelry, and the imaging of faded letters on ancient papyrus, for establishing the provenance of the pieces or in authentication studies [27]. Similarly, x-ray photoelectron spectroscopy (XPS) has been applied to the analysis of archaeological glasses, pigments, metals, and their degradation products.

X-ray microanalysis techniques—in particular, electron probe x-ray microanalysis (EPXMA or EPMA) and SEM coupled with energy dispersive spectrometers (EDS, EDX) are, by far, one of the surface analysis techniques most extensively used in the field of art and art conservation, and they have actually become routine methods of analyzing art and archaeological objects and monitoring conservation treatments [34, 61, 63].

Other surface analysis techniques have been less frequently applied in the analysis of art objects. Auger electron spectroscopy (AES)/scanning Auger microscopy (SAM) in the art and art conservation field involve studies of metallurgical archaeomaterials or the microscopic deteriorated structure of archive and book collections and the effectiveness of the applied conservation methods. Electron-energy-loss spectroscopy (EELS), carried out with the help of a conventional TEM or with a scanning transmission electron microscope (STEM) or Rutherford backscattering spectrometry (RBS), have eventually been applied in the study of objects of artistic and archaeological nature—in particular, characterization of the elemental composition of the wood and protective varnish of old musical instruments, or characterization of metallic archaeological artefacts. Finally, secondary ion mass spectrometry (SIMS) has been applied to the characterization of cross sections of ancient paintings, and in the obsidian hydration dating technique (*vide infra*).

1.4.7 Dating Methods

Dating methods primarily applied in archaeological and archaeometric studies are aimed, in general, at providing the basic chronological framework of a specific region or archaeological site. Despite the diversity of the basic phenomena of dating methods, all follow a general scheme based on a measurable time-dependent quantity or parameter with a well-known value associated with a determined event, which is used as "reference point" or "starting point" in a time-scale relatable to the archaeology of interest [64]. Most of the current dating techniques can be considered absolute, but not all. In general, dating techniques can be classified into different groups according to the type of phenomenon in which they are based as it is overviewed in Scheme 1.10.

(a) Climatic clocks and frameworks. This type of dating technique has evolved from the visual observation of materials of archaeological and geological origin. They are based on the chronological framework provided by the climatic variations taking place on the Earth during the recent geological period (roughly the last two million years) [65].

A study of the loess deposits of windblown sediment that are found in large areas in the middle latitudes of the Northern hemisphere, South America, and New Zealand provide more or less continuous records of the change in the climate in these regions—in particular, the glacial stages [65].

The oxygen-isotope ratio ($^{18}O/^{16}O$) found in the calcareous tests has been used as an indicator of the amount of water locked up in glaciers, which reflects the state of the global climate. Another source of continuous records of oxygen-isotope variation, as well as dust content and acidity, is found in the long cores drilled in polar ice caps in which annual layers can be counted and seasonal variation can be established [65].

The classical work of Milankovitch, further elaborated on by other authors, focuses on the study of the correlation between climate change and the Earth's orbital variations—in particular, the build-up and decay of glaciations [65].

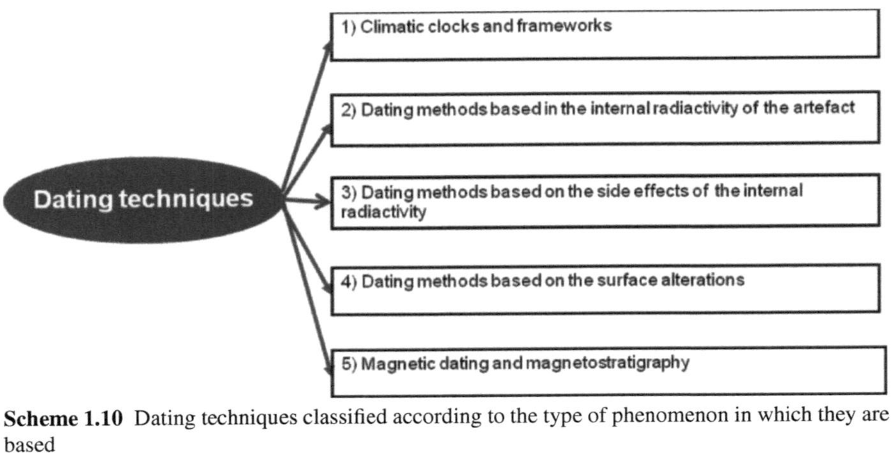

Scheme 1.10 Dating techniques classified according to the type of phenomenon in which they are based

Analysis of pollen and spores preserved in acid soils and peat bogs has given rise to palynology. This approach is based on the assumption that the type of tree and plants growing in a particular terrestrial region are indicator of the climate. Morphological characteristics of the specimens are observed with the help of light and electron microscopes [64].

Varve or rhythmites chronology is another approach based on the examination of geological materials. In this model, the distribution of the laminated sediments formed in the bottom of the beds of dried-out lakes (as a consequence of the seasonal melting of glaciers and the subsequent deposition of coarse particles supplied by streams, followed by finer materials such as sand, silt, and clay) is analyzed and correlated with climate changes [64].

Finally, the dendrochronology first proposed by A. E. Douglas as a dating method has played a relevant role in the development of a general chronological framework. Dendrochronology not only has put a distortion in the prior time-scale into evidence, but also has provided a more accurate calibration establishing a specific relationship between Europe and the Near East during the Neolithic and Bronze Ages [66].

(b) Dating methods based on the internal radioactivity of the artefact. Four dating methods of great relevance in archaeological and archaeometric studies can be included in this section. They can be considered radiometric methods because they are based on the measurement of the internal radioactivity displayed by the object: (a) radiocarbon, (b) potassium-argon, (c) uranium series dating, and (d) lead isotope ratio dating.

Radiocarbon dating originated with W. F. Libby (1955, 1965) and is routinely applied in specialized laboratories of archaeology and in the environmental sciences field. Radiocarbon technique is based on the constant rate at which carbon-14 spontaneously decays. A $^{14}C/^{12}C$ or $^{14}C/^{13}C$ ratio can be determined by measuring the rate of emission of beta particles per unit mass of total carbon, which is proportional to the concentration ratio, and alternatively by measuring the ^{14}C content or the $^{14}C/^{12}C$ or $^{14}C/^{13}C$ ratios in an accelerator mass spectrometer (AMS) [67]. This equipment is not only used for measuring ^{14}C, but also other longer-lived radionuclides such as ^{26}Al, ^{10}Be, ^{36}Cl, ^{41}Ca, and ^{129}I can be determined. This technique has been applied to a wide range of artistic and archaeological objects and human remains [68].

The radioactivity of the ^{40}K isotope, with a half life of 1,250 million years, is used in potassium-argon dating. The experimental determination of the ^{40}Ar content is carried out by MS after releasing the gas via fusion of the sample. Single grains ($\sim$1 mg) can be analyzed by using a high-powered laser to fuse the grain and drive off the argon for measurement in a super-sensitive mass spectrometer. This method is called single crystal laser fusion (SCLF) [68]. Alternatively, determination can be performed by irradiating the sample with neutrons in a nuclear reactor so that a part of the K atoms are converted into ^{39}K. After this, the $^{40}Ar/^{39}Ar$ ratio is measured with a thermal ionization mass spectrometer (TIMS) [68].

The uranium series dating is based on the determination of the $^{234}U/^{230}Th$ or $^{235}U/^{231}Pa$. Isotopic ratio measurements can be made by means of a mass spectrometer. Multicollector inductively coupled plasma mass spectrometry (MC-ICP-MS) or LA-ICP-MS are now used instead of TIMS due to the greatly reduced sample size, increased precision, sampling resolution, and shorter run times that these techniques provide. This technique is mainly applied on stalagmitic calcite formed from ground water where U is present on rock art fossil teeth or coral and mollusk shells [68].

Lead isotopic composition analysis has proven useful for sourcing a number of archaeological materials, including metallic leads, silver, copper, bronze, faience, glass, glazes, and pigments. Determination of lead isotope content can be performed using different methods: TIMS, quadrupole inductively coupled plasma mass spectrometry (Q-ICPMS), inductively coupled plasma sector field mass spectrometry (ICP-SMS), multiple collector inductively coupled plasma mass spectrometry (MC-ICP-MS), and inductively coupled plasma time-of-flight mass spectrometry (ICP-TOF-MS) [69].

(c) Dating methods based on the side effects of internal radioactivity. The methods included in this group are based on the cumulative effect of nuclear radiation on crystal structure. Fission track dating is based on the determination of the ratio of tracks found in a mineral sample (i.e., zircon, obsidian) due to the split of ^{238}U atoms to the tracks induced by exposure in a nuclear reactor. Identification and counting of tracks are performed under a high-magnification microscope from samples prepared as cross sections. Obsidian knife blades and flakes have been dated in the range of 400–600 years. Application to bones and teeth has also been reported [65].

Stimulated luminescence dating methodologies is the term used to refers to two techniques in which the observed signal increases with age. Both thermoluminiscence (TL) and optically stimulated luminescence (OSL) are based on the assumption that many natural samples are able to accumulate (in trap states with a long mean life) electrons that have acquired sufficient energy from α, β, and γ radiations emitted from natural radionuclides belonging to the ^{238}U, ^{232}Th, and ^{40}K decay chains, and from cosmic radiation. Mainly used for dating pottery, they have extended their use to other materials such as polymineral fine-grained fractions from sediments, zircon, calcite and stalagmitic calcite, other salts, meteorites, flint, quartz, loess, volcanic materials (obsidian, tephra), and metallurgical slags [68].

In TL, the light emission induced in the mineral or ceramics sample heated up to 500°C is measured by means of a photomultiplier detector. In addition to laboratory instrumentation, portable gamma spectrometers have been used when circumstances make sampling impractical.

Optical dating techniques, including OSL, optically stimulated phosphorescence (OSP), and phototransferred thermoluminiscence (PTTL) are applied to the dating of anthropological remains such as teeth and bones [65].

ESR has been applied to the dating of secondary carbonates precipitated as stalagtites, stalagmites, and flowstones (speleothems). Other applications include tooth enamel, bone and dentine, and mollusks and shells [64].

(d) Dating methods based on surface alterations. Three methods based on chemical treatments of samples that have undergone surface alterations or isomorphic changes in their composition are included in this category of dating methods:

(a) amino acid racemization, (b) obsidian hydration, and (c) substitution of U, N, or F in bone (FUN dating).

The amino acid racemization method is based on the measurement of the degree of racemization acquired by a material expressed as the ratio "dextro amino acid (D) enantiomeric form/levo amino acid (L) enantiomeric form," which is an indicator of the age of the object. This method has been applied in fossil shells, bones, teeth, wood, plant remains, and coral [68].

Obsidian is a volcanic glass frequently used as tool by prehistoric men. The thickness of the hydration layer that has developed on the surface of the artifact since it was made can be used to date it. Measurement of the hydration layer is carried out by a variety of instrumental techniques. Among them are nuclear resonance reactions, LM, PAS, XPS, XANES, sputter-induced optical spectrometry (SIPS), and SIMS [70].

FUN dating is applied to obtain a relative dating of bones found in the same archaeological site or deposit rather than for absolute dating.

(e) Dating methods based on changes in the direction or orientation of the Earth's magnetic field. Paleomagnetism applies to burnt geological material (volcanic lava) or unburnt sediments that record the polarity changes in the Earth's magnetic field over time. The reversal stratigraphy focuses on the study of Paleomagnetism, providing interesting data for a more accurate establishment of Paleolithic time scales. Parameters such as remanent magnetization or susceptibility are used for this purpose.

Archaeomagnetism can be considered a branch of Paleomagnetism specifically devoted to the dating of archaeological materials from the measurement of the remanent magnetization achieved by the iron oxide impurities present in clay after cooling of the ceramic artifact. This recording mechanism primarily provides information on the direction of the Earth's field at the time the artifact was fired or the kiln was last used.

The main recording mechanism considered in archaeomagnetism is the thermoremanence (TRM). Measurement of this remanent magnetism is experimentally performed by means of spinner magnetometers. This method has been applied in combination with potassium-argon dating for dating remains ascribed to homo erectus (1.64 ± 0.03 million years) or for establishing the age of early man's tools in Asia around two million years ago. Remanent magnetization dating of sunbaked bricks, the discrimination of cast coins from struck ones, or dating the firing temperature of pottery have also been other interesting applications included in archaeomagnetism [71].

1.5 Final Considerations

Electroanalytical techniques and, in particular, solid state electroanalytical techniques introduced in the field of archaeometry, art, and art conservation in recent years (as will be described in the following chapters) provide new and singular information complementary to that of the analytical techniques described in the previous sections. Thus, they not only allow identification of electroactive elements present in the sample, but they can also simultaneously discriminate their state of oxidation. When solid state electroanalytical techniques are coupled with AFM, phase identification can be performed at nanoscale levels on electroactive samples. Additionally, electroanalytical data processed by means of standard addition methods or Tafel analysis can provide, in complex samples containing several pigments, an accurate determination of the content of the different pigments—thus enabling an adequate knowledge of the dosages used by the artist and, as consequence, of its artistic technique. These capabilities situate electroanalytical techniques in a relevant position and give them a promising future as a novel and powerful analytical tool at the service of archaeometry, art, and art conservation.

Chapter 2
Identification of Species by Electrochemical Methods

2.1 Introduction

As previously noted, the identification of components in works of art and archaeological objects is a difficult task during their scientific examination. Samples usually appear as heterogeneous or (apparently) homogeneous solid materials, frequently incorporating different portions; for instance, preparative, pigmenting, and protective layers in paints. At first glance, conventional electrochemical methods are applicable for identifying pigmenting species in solutions resulting from the chemical attack of acids, bases, and organic solvents, etc., of the solid sample. This approach offers two main drawbacks: (i) the disposal of sample amounts restricted to the ng-µm range, thus requiring high sensitive methods, and especially (ii) the loss of analytical information due to the chemical attack on the sample. Thus, an acidic attack of a sample containing either red or yellow cadmium pigments yields Cd^{2+} ions in solution, which can be identified by conventional stripping voltammetry. However, this procedure does not inform on the composition of the original pigment.

Solid state voltammetric methods can be used to obtain information on the composition of the materials used in works of art. Here, the methodology of the voltammetry of microparticles, developed by Scholz et al. [72, 73], will be presented. This methodology provides qualitative, quantitative, and structural information on sparingly soluble solid materials, as described in extensive reviews [74–77] and a precedent monograph [78], just requiring sample amounts in the ng-µg level.

2.2 Conventional Voltammetry

Although a variety of electrochemical methods are available, voltammetry (based on the record of the current i flowing across an electrochemical cell under the application of a given time-depending potential E) has become the most important measuring technique in pure and applied electrochemistry [79].

Generally, a voltage is applied between a working electrode and an auxiliary electrode, and the potential of the working electrode is controlled versus the reference electrode possessing a constant and known potential. The current flowing through

A. Doménech-Carbó et al., *Electrochemical Methods in Archaeometry, Conservation and Restoration*, Monographs in Electrochemistry,
DOI 10.1007/978-3-540-92868-3_2, © Springer-Verlag Berlin Heidelberg 2009

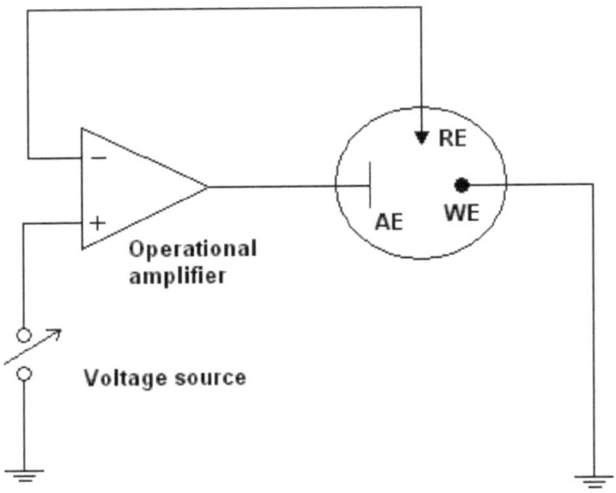

Fig. 2.1 Scheme for a conventional three-electrode voltammetric cell. RE = reference electrode; WE = working electrode; and AE = auxiliary electrode

the working and auxiliary electrodes is the response to the potential of the working electrode—i.e., the excitation signal. Usually a three-electrode arrangement is used, as schematized in Fig. 2.1. In this equipment, a potentiostat controls the potential so that the current flows almost exclusively between the working electrode and the auxiliary electrode, while a very small, practically negligible current is passing through the reference electrode—usually AgCl/Ag, or Hg_2Cl_2/Hg (calomel) electrodes.

Depending on the time variation of the applied potential, several types of voltammetry can be distinguished. Among them, the most widely used are linear and cyclic voltammetries. Here, the excitation signal is a linear potential scan that is swept between two extreme values, and in cyclic voltammetry the potential is swept up and down between the two values (or switching potentials) with the same absolute scan rate (v, usually expressed in mV/s), although it has the opposite sign [79].

In conventional cyclic and linear sweep voltammetry, electrochemical measurements are performed in a quiescent solution of an electroactive species contained in a given electrolyte. The concentration of the electroactive species is typically between 10 μM and 1 mM, while a supporting electrolyte—typically in 0.10 M concentration—is added. The resulting solution must generally be deoxygenated to avoid perturbing signals and possible secondary reactions with dissolved oxygen. Purging the solution with nitrogen or argon for 5–15 min is performed for this purpose.

Figure 2.2 shows the first-scan cyclic voltammogram obtained by a platinum electrode for a 1.0 mM $K_4[Fe(CN)_6]$ plus 0.10 M KCl solution in water. Here, as in all figures in this chapter, potentials are referred to the AgCl (3 M NaCl)/Ag electrode. Generally, the first recorded cycles differ slightly, and only after 3–4 cycles do the CVs approach a constant response. In this experiment, the potential scan is

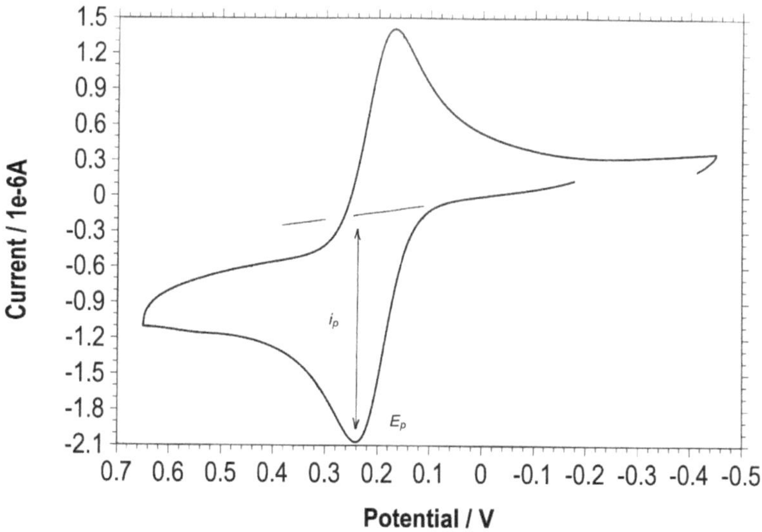

Fig. 2.2 Cyclic voltammogram at Pt electrode for a 1.0 mM $K_4Fe(CN)_6$, plus 0.10 M KCl solution. Potential scan rate of 50 mV/s. The peak current measured from the base line and the peak current for the anodic peak are marked

initiated at -0.18 V in the positive direction, and switches at $+0.65$ V, thus defining an initial anodic (oxidative) scan. In the subsequent scan, the potential is negatively shifted, defining a cathodic (reductive) scan until the new switching potential of -0.45 V is reached. In this system, a redox process based on the oxidation of iron ions in their oxidation state $+2$ to iron ions in a $+3$ oxidation state can occur. This defines a $[Fe(CN)_6]^{3-}/[Fe(CN)_6]^{4-}$ redox couple, characterized by a mid-peak potential E_{mp} that can, under certain conditions, be taken as a good approximation for the formal electrode potential $E_c^{\circ\prime}$ of the redox system—in turn, directly related to the thermodynamic electrode potential E°.

Initially, at potentials between -0.25 and about $+0.1$ V, a rather low current is recorded due to double-layer charging. Later, when the potential is sufficiently positive to oxidized $[Fe(CN)_6]^{4-}$ to $[Fe(CN)_6]^{3-}$, the anodic current increases rapidly until the concentration of $[Fe(CN)_6]^{4-}$ at the electrode surface is substantially diminished. Then, the (anodic) current increases dramatically until a maximum value is reached, thus defining a (anodic) voltammetric peak with the peak potential E_{pa} and the peak current i_{pa}. The correct peak current must be measured in relation to the background current, which can be extrapolated from the starting region. After the peak, the current decreases slowly as the electrode is depleted of $[Fe(CN)_6]^{4-}$ due to its electrochemical conversion into $[Fe(CN)_6]^{3-}$. When the (anodic) switching potential $E_{\lambda a}$ is attained, the potential scan reverses its direction. In the subsequent cathodic scan, a similar cathodic peak is measured, defining a cathodic peak potential E_{pc} and a cathodic peak current i_{pc}. Then, the current reaches a maximum and subsequently decays.

This system can be taken as an example of so-called reversible, diffusion-controlled electrochemical processes. In short, during the initial anodic scan, an electron-transfer process between the ferrocyanide (or hexacyanoferrate(II)) ions, $[Fe(CN)_6]^{4-}$, and the working electrode occurs. This can be represented by means of the equation (here, aq denotes species in aqueous solution):

$$[Fe(CN)_6]^{4-}_{aq} \rightarrow [Fe(CN)_6]^{3-}_{aq} + e^- \tag{2.1}$$

This process can be described in terms of a heterogeneous reaction in which ferricyanide (or hexacyanoferrate(III)) ions, $[Fe(CN)_6]^{3-}$, are formed. At the beginning of the voltammetric peak, the current is controlled by the kinetic of the electron transfer across the electrode/electrolyte barrier so that the current increases somewhat exponentially with the applied potential. The value of the current is controlled 150–200 mV after the voltammetric peak by the diffusion rate of ferrocyanide ions from the solution bulk toward the electrode surface.

In the reverse (cathodic) scan, ferricyanide ions remaining in the vicinity of the electrode surface are reduced to ferrocyanide ones. This process can be represented by the inverse of Eq. 2.1, with the voltammetric profile being interpreted from considerations similar to those made for the anodic peak.

The theory for cyclic voltammetry was developed by Nicholson and Shain [80]. The mid-peak potential of the anodic and cathodic peak potentials obtained under our experimental conditions defines an electrolyte-dependent formal electrode potential for the $[Fe(CN)_6]^{3-}/[Fe(CN)_6]^{4-}$ couple $E^{\circ\prime}$, whose meaning is close to the genuine thermodynamic, electrolyte-independent, electrode potential E° [79, 80]. For electrochemically reversible systems, the value of $E^{\circ\prime}$ $(= (E_{pc} + E_{pa})/2)$ remains constant upon varying the potential scan rate, while the peak potential separation provides information on the number of electrons involved in the electrochemical process: $(E_{pa} - E_{pc}) = 59/\mathrm{n}$ mV at 298 K [79, 80]. Another interesting relationship is provided by the variation of peak current on the potential scan rate: for diffusion-controlled processes, i_p becomes proportional to the square root of the potential scan rate, while in the case of reactants confined to the electrode surface, i_p is proportional to v [79].

In voltammetric experiments, electroactive species in solution are transported to the surface of the electrodes where they undergo charge transfer processes. In the most simple of cases, electron-transfer processes behave reversibly, and diffusion in solution acts as a rate-determining step. However, in most cases, the voltammetric pattern becomes more complicated. The main reasons for causing deviations from reversible behavior include: (i) a slow kinetics of interfacial electron transfer, (ii) the presence of parallel chemical reactions in the solution phase, (iii) and the occurrence of surface effects such as gas evolution and/or adsorption/desorption and/or formation/dissolution of solid deposits. Further, voltammetric curves can be distorted by uncompensated ohmic drops and capacitive effects in the cell [81–83].

Corresponding to a cyclic voltammogram (CV), recorded at a glassy carbon electrode (GCE) immersed into a 1.0 mM Cu^{2+} solution in acetic acid/sodium acetate

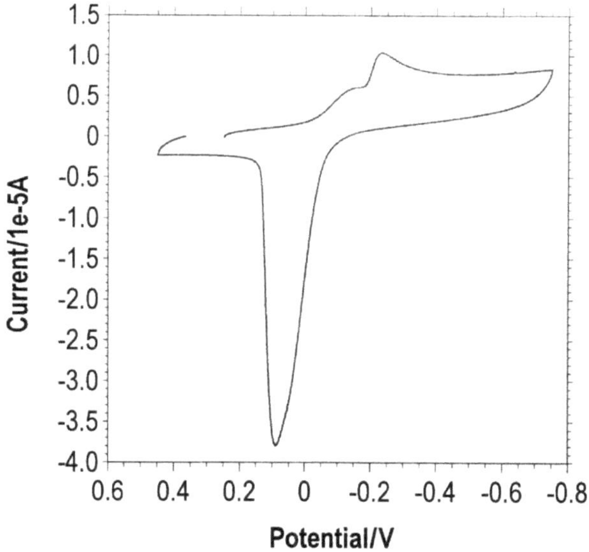

Fig. 2.3 Cyclic voltammogram at GCE for a 1.0 mM solution of $CuSO_4 \cdot 5H_2O$ in 0.25 M HAc + 0.25 M NaAc, pH 4.85. Potential scan rate 50 mV/s

buffer, Fig. 2.3 allows for an examination of some of these processes. Here, the potential is initially scanned in a negative direction so that a main reduction peak appears at −0.22 V. In the subsequent anodic scan, a sharp oxidation peak is recorded at +0.08 V. The initial two-electron reduction of copper ions can be represented as:

$$Cu^{2+}{}_{aq} + 2e^- \rightarrow Cu_s \tag{2.2}$$

As a result of that reductive process, a deposit of copper metal (denoted in Eq. 2.2 by "s" for "solid") is formed on the carbon electrode surface. The prominent anodic peak recorded in the reverse scan corresponds to the oxidative dissolution of the deposit of copper metal previously formed. The reason for the very intense anodic peak current is that the copper deposit is dissolved in a very small time range (i.e., potential range) because, in the dissolution of the thin copper layer, practically no diffusion limitations are involved, whereas in the deposition process (i.e., the cathodic peak), the copper ions have to diffuse through the expanding diffusion layer from the solution to the electrode surface. These processes, labeled as stripping processes, are typical of electrochemically deposited metals such as cadmium, copper, lead, mercury, zinc, etc., and are used for trace analysis in solution [84]. Remarkably, the peak profile is rather symmetrical because no solution-like diffusive behavior is observed.

For analytical purposes, voltammetric peaks can be used for identifying species in a solution phase, provided that a well-defined electrochemical response is obtained. Application of that methodology implies a sample treatment is needed to dissolve electroactive components. Examples involve extraction of organic dyes in fibers and inks by organic solvents or attack of metallic fragments by mineral acids.

The use of analytical procedures based on conventional voltammetry requires the selection of a solvent, a supporting electrolyte, an electrode, an electrochemical technique, electrochemical parameters, and data treatment:

(a) Solvent selection derives from a compromise between the desired solubility of the analyte(s) and the available potential range. The solvent, electrolyte, and electrode material determine the potential window of operative potentials. Thus, in neutral aqueous solutions with graphite electrodes, large cathodic currents appear due to water reduction into hydrogen at about −1.25 V vs. saturated calomel electrode (SCE), while large anodic currents are recorded above +1.25 V due to water oxidation into oxygen. With gold and platinum electrodes, hydrogen discharge is favored, and the available potential range is restricted to the potential region between −1.00 and +1.25 V.

(b) In organic solvents such as CH_3CN (MeCN), $(CH_3)_2SO$ (DMSO), HCON $(CH_3)_2$ (DMF) or CH_2Cl_2, the salts Et_4NClO_4 and Bu_4NPF_6 (among others) are the most used supporting electrolytes. In aqueous solutions, a variety of supporting electrolytes can be used, from salts (NaCl, $NaClO_4$) to acids (HCl, H_2SO_4, HNO_3) to alkalis (NaOH,KOH). The use of buffers such as acetate and phosphate in concentrations of 0.10–0.50 M allows for reproducible experimental conditions. In some instances, complexing agents such as EDTA, oxalate, and tartrate can be added.

(c) The electrode material can seriously condition the observed response, as previously noted. Apart from solvent discharge effects, electrodes are to always "active" to some extent. Thus, in contact with aqueous electrolytes, mercury electrodes are oxidized to Hg_2^{2+} and Hg^{2+} at potentials above +0.20 V vs. AgCl/Ag, while gold and platinum electrodes form surface oxides at potentials ca. +1.0 V.

(d) Linear potential scan and cyclic voltammetries (LSV and CV, respectively) are the most widely used electrochemical techniques. Of course, nonvoltammetric techniques, such as controlled potential and controlled current coulometry, chronoamperometry, and chronopotentiometry, can also be used. Within the context of voltammetric methods, other techniques are available using a suitable function generator. In differential pulse voltammetry (DPV), potential pulses are superimposed to a linear potential scan [79]. In square wave voltammetry (SQWV), a square wave signal comprised of the sum of a symmetrical square wave of half peak-to-peak amplitude E_{SW} and a staircase of the same period τ is used. The potential increment of the staircase is ΔE. The current is sampled at the end of each pulse to yield forward and reverse currents. Their difference is the net current. These above techniques minimize capacitive effects and yields tall peaks, thus facilitating resolution in systems displaying multiple signals [85].

(e) Electrochemical parameters can be varied within relatively large intervals, thus modulating the observed electrochemical response. In cyclic voltammetry, increasing potential scan rate generally results in increasing peak currents, but in a concomitant increase of background currents (and shape-distorting capacitive and resistive effects, in particular), as well as in a decrease of peak resolution. In pulse techniques, pulse amplitude, pulse width, and pulse intervals can be varied, resulting in different conditions for studying electrochemical processes. Variation of such parameters allows performance of voltammetric experiments in "short-time" and "long-time" experiments with experimentation times between 0.1–1 ms and 10–100 min, respectively.

(f) Current electrochemical equipment routinely provides computer-assisted treatment of data, determining peak current and peak potential, peak area, and displaying derivative and integral curves. Derivative and integral techniques can be used for improving peak resolution. In particular, semiderivation (deconvolution) and semi-integration (convolution) techniques are frequently used [86].

This is illustrated in Fig. 2.4, where the deconvolution of differential pulse voltammograms at the glassy carbon electrode in the ethanol extract from two commercial inks are shown. Samples were taken from paper fragments of 0.10 mg immersed for 10 min in a 50:50 (v/v) ethanol: 0.50 M aqueous acetate buffer (pH 4.85) solution.

Since low volumes of solution and low analyte concentrations are involved in samples from works of art and archaeological pieces, low-volume cells (1–2 mL) can also be used coupled with micro- and ultramicroelectrodes. These last provide high sensitivity and allow operation with electrolytes of low dielectric constant or even work without supporting electrolytes [87, 88].

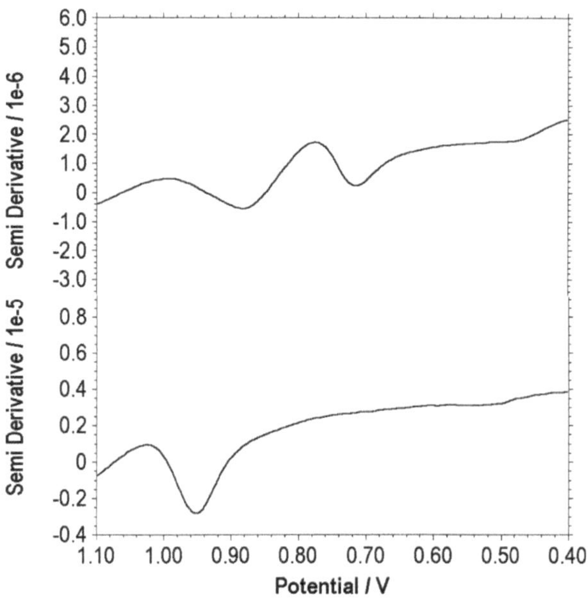

Fig. 2.4 Deconvolution of differential pulse voltammograms at the glassy carbon electrode in the ethanol extract from two commercial inks transferred to a 50:50 (v/v) ethanol:0.50 M aqueous acetate buffer (pH 4.85) solutions. Voltammograms for red (upper) and blue (below) commerical inks

2.3 Voltammetry of Microparticles

The application of solution phase electrochemistry for studying the work of art samples and archaeological artifacts requires, as previously indicated, sample treatment via extraction or chemical attack. This is an obvious drawback, because these operations, apart from the requirement of relatively high amounts of sample, usually lead to a loss of information since the solid is dissolved and all signals that are specific for the solid compound or material are not available anymore for measurements.

The first use of electrochemical methods for identifying components in solid samples can be attributed to Fritz [89] and Glazunov [90], who developed the electrography. Here, two inert electrodes are connected to a potential source, and one is brought into contact with the solid sample (mineral fragment, coin, metallic artifact), which is placed with one rim in contact with a filter paper moistened with a selected electrolyte. The second electrode is connected with the filter paper. In the most common version, the current is allowed to pass for a few seconds so that metallic constituents experience an anodic dissolution. A related procedure was the Weisz ring-oven method [91], further developed by Stephen [92], with direct application to the analysis of coins. Identification and semiquantitation of metal ions can be performed colorimetrically by adding the adequate reactants to the electrolyte, which impregnates the filter paper (diethyldithiocarbamate, demiethylglyoxime, $K_4Fe(CN)_6$, etc.). As a result, both qualitative [93–95] and semiquantitative [96] analysis of metal cations from the solid sample have been reported.

Carbon paste electrodes (CPEs), introduced in the 1960s by Kuwana et al. [97, 98], and further studied by Bauer et al. [99–101] and Brainina et al. [102, 103], were the most widely used approach for studying the electrochemical response of solids. Here, the solid is dispersed into a conducting paste formed by graphite powder and an organic (nujol oil, paraffin oil, etc.) or inorganic (aqueous electrolytes, such as H_2SO_4) binder. Although CPEs have produced a relatively extensive literature focused on mineral analysis using both organic [104] and inorganic [105] binders, their application to pigment analysis suffers from the presence of large background currents and the high sensitivity of the electrochemical response to changes in the composition of the binder. In 1989, Scholz et al. [72, 73] introduced the voltammetry of microparticles (also called abrasive stripping voltammetry, when the transferred solid is electrochemically dissolved during the measurement), where the solid sample is transferred by abrasion to the surface of paraffin-impregnated graphite electrodes (PIGEs). This methodology provides large signal-to-background responses and excellent repeatability, allowing for identification of inorganic [106] and organic [107] pigments. Alternative approaches involve graphite-polymer composites [108], formation of fine deposits of the solid sample by evaporation of solutions [109] or suspensions [110] of the solid on conventional electrodes, and attachment of solid microparticles to polymer coatings (polymer film electrodes, PFEs) [111, 112]. Selection of the type of electrode modification may be crucial for obtaining an electrochemical response that is useful for analytical purposes.

Thus, the study of the electrochemistry of nonconducting solid materials requires their "contact" with a substrate electrode or their dispersion into a conducting matrix. In typical carbon paste electrodes, a small amount of the solid analyte is mixed with graphite powder plus a nonelectrolyte binder to form a conducting past [97, 98, 104]. Then, the electron transfer processes occur at the paste-electrolyte interface, the currents are smaller, and the surface coverage can be minimized. However, the particle size distribution, the nature of the binder, and paste composition influence the electrochemical response, which frequently exhibits poor reproducibility. These problems can be solved in part by using an electrolyte binder that ensures large conductivity within the paste [99–101, 105, 113, 114]. However, aggressive binders may dissolve the solid analyte or produce undesirable chemical reactions.

These drawbacks can be avoided to a large extent, using the voltammetry of microparticles—a technique involving solid state electrochemistry where down to about 10^{-6} to 10^{-11} mol of sample [74–78] can be transferred by abrasion into the surface of an inert electrode, usually paraffin-impregnated graphite electrodes, and the electrode is later immersed in a suitable electrolyte for recording its voltammetric response. The response of this sample-modified electrode, consisting of the reduction or oxidation of the solid materials, becomes phase-characteristic.

As in solution phase electrochemistry, selection of solvent and supporting electrolytes, electrode material, and method of electrode modification, electrochemical technique, parameters and data treatment, is required. In general, long-time voltammetric experiments will be preferred because solid state electrochemical processes involve diffusion and surface reactions whose typical rates are lower than those involved in solution phase electrochemistry.

A paradigmatic example is provided by an extensively used pigment: Prussian blue— potassium-iron(III) hexacyanoferrate(II), also called potassium-iron(III) ferrocyanide. This is constituted by potassium-iron(III) ferrocyanide, $KFe[Fe(CN)_6]$, whose electrochemistry has been extensively studied [74–77].

As shown in Fig. 2.5, the cyclic voltammograms for Prussian blue attached to paraffin-impregnated graphite electrodes (PIGEs) in contact with aqueous electrolytes exhibit two well-defined one-electron couples. Prussian blue crystals possess a cubic structure, with carbon-coordinated Fe^{2+} ions and nitrogen-coordinated Fe^{3+} ions, in which potassium ions, and eventually some Fe^{3+} ions, are placed in the holes of the cubes as interstitial ions. The redox couple at more positive potentials can be described as a solid-state process involving the oxidation of Fe^{2+} ions. Charge conservation requires the parallel expulsion of K^+ ions [77]:

$$KFe^{III}[Fe^{II}(CN)_6]_s \rightarrow Fe^{III}[Fe^{III}(CN)_6]_s + K^+{}_{aq} + e^- \tag{2.3}$$

The process at more negative potentials must correspond to the reduction of Fe^{3+} ions coupled with the insertion of K^+ ions from the electrolyte:

$$KFe^{III}[Fe^{II}(CN)_6]_s + K^+_{aq} + e^- \rightarrow K_2Fe^{II}[Fe^{II}(CN)_6]_s \tag{2.4}$$

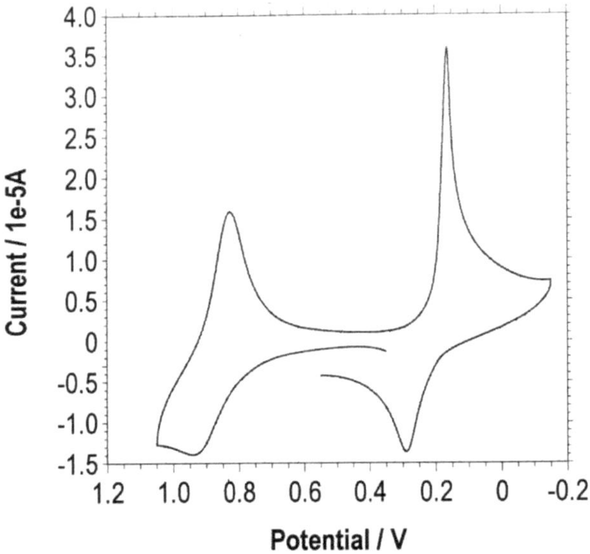

Fig. 2.5 Cyclic voltammogram for a Prussian-blue modified PIGE immersed in 0.10 M KCl. Potential scan rate 10 mV/s

Such electrochemical processes can be described on the basis of the model developed by Lovric and Scholz [115, 116] and Oldham [117] for the redox reactivity of nonconducting solids able to be permeated by cations or anions (so-called ion-insertion solids). As described in the most recent version of Schröder et al. [118], the electrochemical process is initiated at the three-phase junction between the electrode, the electrolyte solution, and the solid particle, as schematized in Fig. 2.6. From this point, the reaction expands via charge diffusion across the solid particle. It is assumed that, for a reduction process, there is a flux of electrons through the

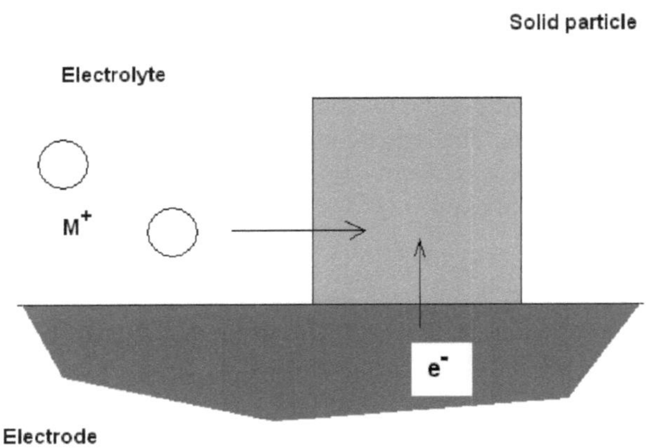

Fig. 2.6 Schematic diagram for describing electrochemical processes (reduction) on a nonconducting solid, allowing for both electron and ion transport

electrode/particle interface, coupled with a flux of electrolyte ions across the electrolyte/particle interface. The electrochemical reaction progresses via electron hopping between immobile redox centers in the solid, accompanied by the migration of charge-balancing electrolyte cations through the solid (electron diffusion and cation diffusion, respectively). A pictorial representation of the described mechanism is illustrated in Fig. 2.6. Obviously, an equivalent description is eventually obtained if the insertion of cations into the solid is replaced by the issue of mobile anions from the solid to the electrolyte.

The relevant point to emphasize is that this model allows one to justify the existence of solid-state electrochemical reactions. Remarkably, the redox conductivity is maintained even in the limiting cases where electron diffusion or cation diffusion are hindered. Here, the electrochemical reaction may progress via surface diffusion along the external layer of the particle in contact, respectively, with the electrode and the electrolyte.

The foregoing considerations can also be applied to the electrochemistry of a number of organic compounds in contact with aqueous buffers [107, 119–125]. Here, protonation/deprotonation reactions are coupled with electron transfer processes, as described for the case of indigoid-, anthraquinonic-, and flavonoid-type pigments, among others. In contact with aqueous electrolytes, the electrochemical processes can generally be described as:

$$A^{ox}{}_{s} + qH^{+}{}_{aq} + qe^{-} \rightarrow H_qA^{rd}{}_{s} \tag{2.5}$$

where A^{ox} represents the oxidated form of the organic species and H_qA^{rd} represents the q-protonated reduced form. Figure 2.7 shows a pictorial representation of such electrochemical processes.

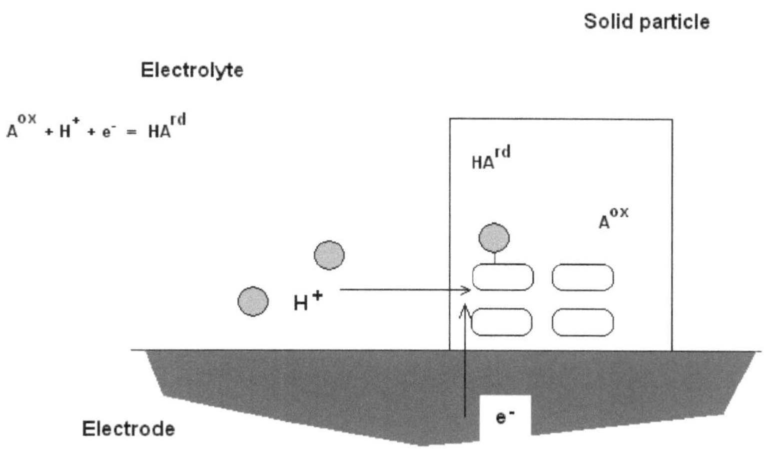

Fig. 2.7 Schematic representation of electrochemical processes (reduction) for organic solids able to experience coupled proton transport/electron transport processes

2.4 Identification of Species Involving Electrochemically Depositable Metals

An important class of solid state electrochemical reactions involves the reduction of metal oxides or salts to the corresponding metals. This situation applies for cadmium, cobalt, copper, lead, mercury, and zinc pigments, whose electrochemical identification using the voltammetry of microparticles approach has recently been discussed for pictorial samples [108, 111, 126, 127] and ceramic pigmenting species [112, 128, 129]. In Fig. 2.8, the cyclic voltammogram is obtained from a $PbCO_3$-modified PIGE immersed into an acetate buffer. This response can be described in the following terms. In the initial cathodic scan, $PbCO_3$ is reduced at -0.60 V to lead metal, the overall process being

$$PbCO_{3s} + 2H^+_{aq} + 2e^- \rightarrow Pb_s + CO_2 + H_2O \quad (2.6)$$

As before, the subscript s denotes solid phases, and the subscript aq denotes solvated ions in solution. As a result of this process, a deposit of lead metal is formed on the electrode surface at the end of the cathodic scan while CO_2 diffuses to the

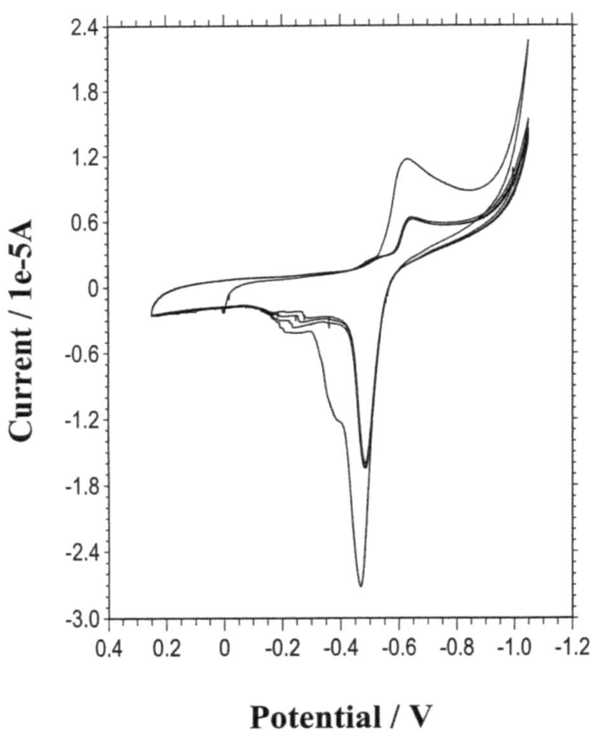

Fig. 2.8 Cyclic voltammogram for a $PbCO_3$-modified PIGE immersed into 0.50 M sodium acetate buffer at pH 4.85. Potential scan rate 50 mV/s

solution bulk. In the subsequent anodic scan, the metallic lead is oxidized to Pb^{2+} ions in solution via an oxidative dissolution process labeled as "oxidative stripping" at −0.48 V:

$$Pb_s \rightarrow Pb^{2+}{}_{aq} + 2e^- \qquad (2.7)$$

Sharp stripping peaks produce large peak currents. In fact, stripping peaks are systematically used in trace analysis for metal ion determination in solution [84].

Upon repetitive cycling of the potential scan, the voltammetric record is reproduced, but an additional cathodic peak near to −0.45 V appears. This is due to the reduction of Pb^{2+} ions electrochemically generated by the previous oxidation of lead metal. The reduction of lead ions occurs at a potential different from that at which the reduction of $PbCO_3$ takes place. In repetitive voltammetry, additional anodic peaks appear at −0.40 and −0.28 V. These are due to the oxidative dissolution of different lead deposits generated in reductive scans [130, 131].

Formation of lead deposits can be monitored by coupling atomic force microscopy (AFM) with electrochemical experiments. Figure 2.9 shows AFM images recorded for Naples yellow (lead antimoniate, $Pb(AsO_3)_2$) deposited onto a carbon plate in contact with acetate buffer. Here, deposits of lead metal appear as sharp grains in the vicinity of the aggregates of the pigment. This situation is similar to that described for litharge (yellow form of PbO) by Scholz et al. [130, 132]. Here, in situ

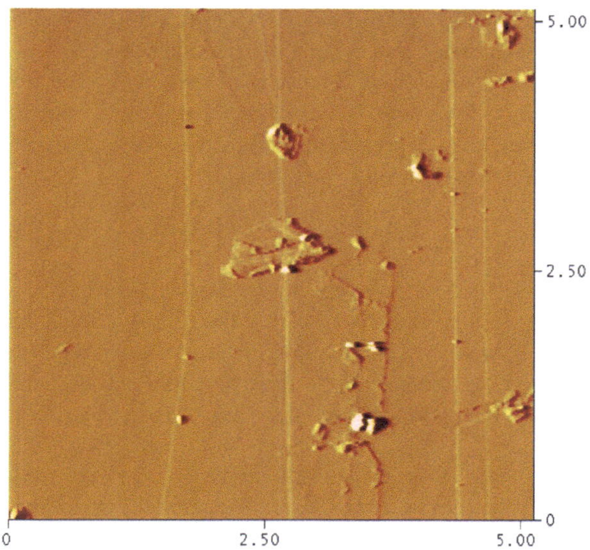

Fig. 2.9 AFM image of a deposit of Naples yellow on a carbon plate in contact with 0.20 M HAc + 0.25 M NaAc at pH 4.85, after application of a potential step of −0.75 V over 2 min [126]

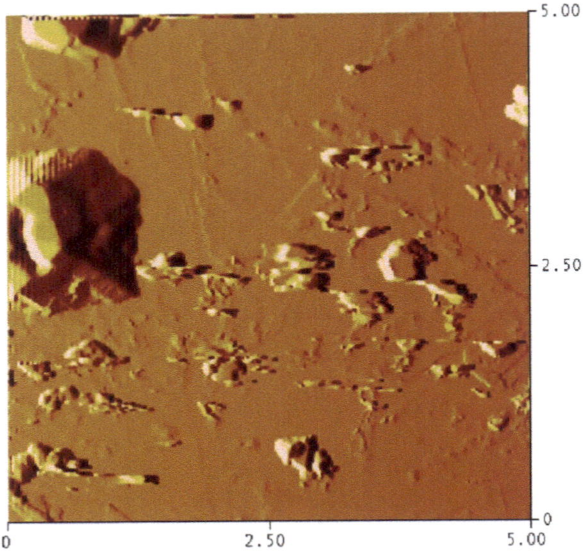

Fig. 2.10 AFM image of a deposit of minium on a carbon plate in contact with 0.20 M HAc + 0.25 M NaAc at pH 4.85, after application of a potential step of −0.75 V over 2 min [126]

XRD [132] and AFM [130] data suggests that the electrochemical mechanism involves a solid state epitactic transformation of lead oxide to lead metal.

In some cases, the deposit of lead metal also consist of irregular aggregates far from the original particles of the lead compound [126], as can be seen in Fig. 2.10 for minium (Pb_3O_4)—another widely used pigment These features can be rationalized, assuming that the electrochemical pathway represented by Eq. 2.6 can eventually be coupled by a reductive process via formation of Pb^{2+} in solution

$$PbO_s + 2H^+{}_{aq} + 2xe^- \rightarrow (1-x)Pb^{2+}{}_{aq} \cdots + xPb_s + H_2O \tag{2.8}$$

further reduced to lead metal via the process inverse to that described by Eq. 2.7.

For our purposes, the relevant point to emphasize is that the different lead pigments yield different reductive responses [126]. Detailed electrochemical analysis indicated that, for a given electrolyte and material, the peak potential and the morphology of the cathodic signals depend slightly on the shape and size distribution of solid particles [73–78].

The shape and position of the anodic stripping peak should be, in principle, independent of the starting material. This peak, however, is influenced by the conditions of metal deposition. Thus, as described by Komorsky-Lovric et al. [131], the stripping peak of electrochemically deposited lead differs from that recorded for lead compounds mechanically attached to graphite electrodes. Additionally, the presence of other depositable metals often gives rise to intermetallic compounds whose subsequent oxidative dissolution differs from those recorded for the individual metals separately [84]. In spite of these drawbacks, anodic stripping peaks can be used for a very sensitive metal detection because their height can be increased by following an analytical strategy well-known in conventional stripping analysis [84]. This consists of the application of a constant potential sufficiently negative to ensure the reduction of the starting compound to the corresponding metal. This electrodeposition (or electrogeneration) step increases the amount of the deposited metal and, consequently, enhances the peak current of the oxidative dissolution process when the potential is again scanned in the positive direction.

The usefulness of this kind of analysis is illustrated in Fig. 2.11 where the square wave voltammograms for azurite, smalt, and two samples from the frescoes painted by Palomino (1707) in the Sant Joan del Mercat church in Valencia (Spain) are shown. Copper pigments yield a unique stripping peak at −0.05 V, whereas cobalt pigments produce a main anodic peak at +0.02 V, accompanied by overlapping

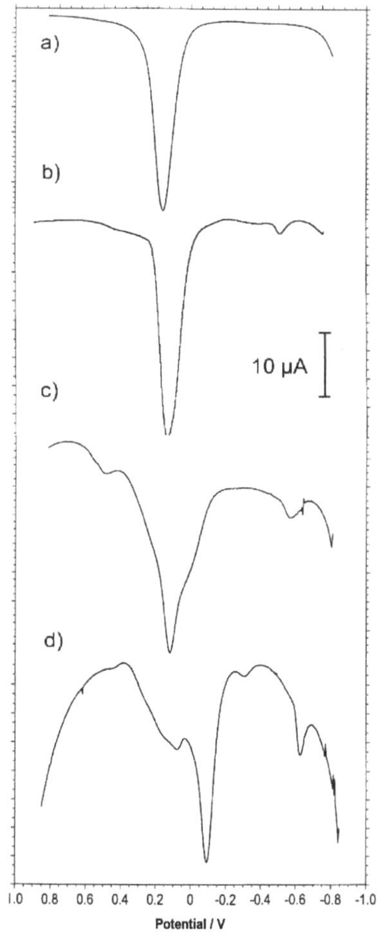

Fig. 2.11 Square wave voltammograms for blanks of azurite and smalt, and for two samples from the Palomino's frescoes in the *Sant Joan del Mercat* church in Valencia, in contact with 0.50 M potassium phosphate buffer, pH 7.4: (**a**) azurite, (**b**) sample PV8b, (**c**) smalt, and (**d**) sample PA5b. Potential scan initiated at −0.85 mV in the positive direction. Potential step increment 4 mV; square wave amplitude 15 mV; frequency 2 Hz [133]

peaks at -0.02 and $+0.22$ V. One of the samples yields only the stripping peak for copper, while the other exhibits a more complicated profile that can be described as the result of the superposition of stripping peaks for copper and cobalt. Interestingly, in this voltammogram, an additional stripping peak appears at -0.48 V, due to the presence of Naples yellow in the sample [126, 133]. Similar considerations can be applied for elucidating the presence of cinnabar and minium in pictorial samples [111]. It is known that cinnabar, a relatively expensive pigment, was frequently used by the artists in "noble" red portions of their paints, replacing it by minium and minium/cinnabar mixtures in other portions. Electrochemical data provide easy and sensitive proof of this practice during the Middle Age.

2.5 Identification of Metallic Species

Identification of metal particles dispersed in pictorial or decorative layers in works of art is often difficult for microscopy techniques because of: (i) their presence as highly diluted components and concentrated in microparticles that, in turn, are included in binding media and attached to priming and protective layers; (ii) the coexistence of metals with priming, ground or pigmenting layers in the samples; and (iii) the presence of products resulting from the alteration of metals.

It is therefore frequently difficult to find punctual areas in the sample having a sufficient concentration of the desired analyte to be detected by the x-ray microanalysis system. Thus, identification and eventually quantitation of metals in decorative/protective layers of pictorial samples by SEM/EDX frequently require an accurate and often time-consuming scanning process.

Metal particles display highly sensitive oxidative dissolution processes resulting in very typical stripping peaks in voltammograms. This is illustrated in Fig. 2.12, where the square wave voltammetric response of a gold-containing sample from the vault *Sant Joan del Mercat* church in Valencia (Spain) is depicted. This frescoe was painted by Acisclo Palomino in 1707 and suffered considerable damage from gunfire during the Spanish civil war in 1936. As a result, strongly deteriorated paint layers survived where the presence of gold microparticles was not clearly detected by SEM/EDX because of the large dilution of such particles in the pictorial layer. Using acetate buffer plus NaCl, the voltammogram exhibits gold-characteristic anodic peaks near to -0.2 and $+1.1$ V, preceded by the stripping of lead at -0.48 V due to the presence of lead pigments (Naples yellow and lead white). A similar situation was described of gold traces in the *Sant Giovanni* church in Taormina (Italy) [134]. The oxidation processes at the more positive potential can, in principle, be represented as [135, 136]

$$2Au_s + 3H_2O \rightarrow Au_2O_{3s} + 6H^+_{aq} + 6e^- \tag{2.9}$$

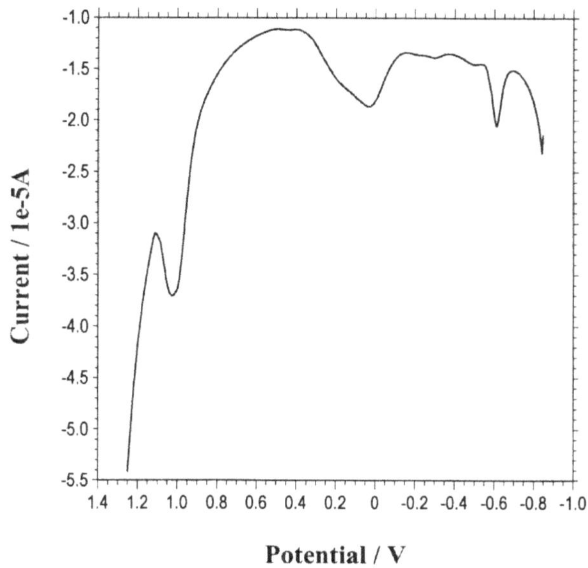

Fig. 2.12 Square wave voltammogram for a sample from the ceiling vault of the *Sant Joan del Mercat* church in Valencia (Spain), painted by Aciclo Palomino in 1707. Electrolyte: 0.50 M HAc + 0.50 M NaAc aqueous buffer plus 0.05 M NaCl (pH 4.85). Potential scan initiated at −850 mV in the positive direction. Potential step increment 4 mV; square wave amplitude 25 mV; frequency 5 Hz

while overlapping processes at less positive potentials may be attributed to the formation of gold-chloride complexes in the solution phase [134]:

$$Au_s + 4Cl^-_{aq} \rightarrow AuCl^-_{4\ aq} + 3e^- \quad (2.10)$$

2.6 Identification of Species Using Reductive/Oxidative Dissolution Process

A number of electrochemical processes involving solid materials can be described in terms of the reductive or oxidative dissolution of such materials. Within this type of processes, one can include the stripping of metal deposits previously mentioned. In the context of archaeometry, conservation, and restoration sciences, the reductive dissolution of iron oxide-type materials is of particular interest. Thus, application of the voltammetry of the microparticles approach for identifying iron pigments has been described [108, 137–139].

A typical example of these processes is shown in Fig. 2.13, where the cyclic voltammetry of (a) hematite, and (b) goethite-modified PIGEs immersed into 0.10 M HCl are depicted. As can be seen in this figure, a prominent cathodic peak is

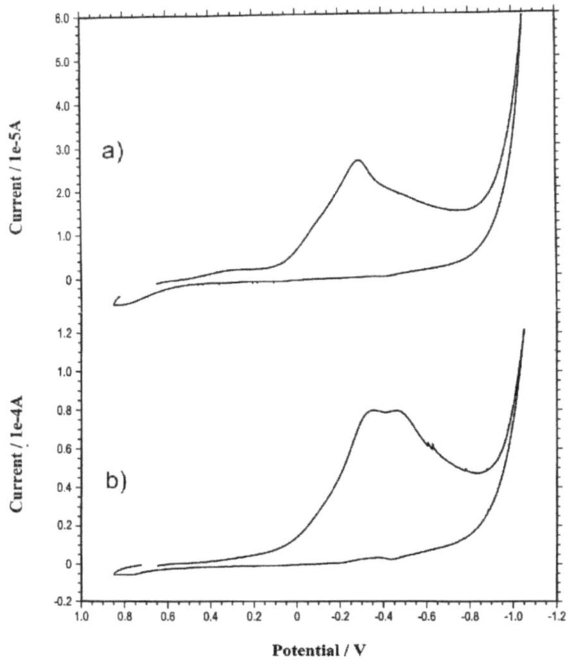

Fig. 2.13 CVs of of PIGEs modified with: (**a**) hematite, (**b**) goethite immersed into 0.10 M HCl. Potential scan rate 50 mV/s [138]

recorded at −0.45 V, with no coupled anodic signals. Interestingly, the voltammetric peaks exhibit a sharp current decay after the peak, clearly differing from the smooth decays typically observed in solution-phase electrochemistry (see Fig. 2.2). The shape of voltammetric curves depends on the mineralogical composition, but also on the shape and size distribution, of particles. Then, voltammograms of different pigments provide remarkably different profiles, as shown in Fig. 2.14. Here, square wave voltammograms for (a) *Caput mortum*, and (b) Venetian red attached to PIGEs in contact with 0.10 M HCl are depicted. The electrochemistry of iron oxides and related minerals has attracted considerable attention [137–148]. Following Grygar [144–148], this voltammetry can be described in terms of the reductive dissolution of solid Fe_2O_3 (hematite) and FeO(OH) (Goethite) to Fe^{2+} ions in solution:

$$Fe_2O_{3s} + 6H^+_{aq} + 2e^- \rightarrow 2Fe^{2+}_{aq} + 3H_2O \tag{2.11}$$

$$FeO(OH)_s + 3H^+_{aq} + e^- \rightarrow Fe^{2+}_{aq} + 2H_2O \tag{2.12}$$

Oxidative dissolution processes can also be used for identifying solid materials. This is the case of chromium (III) oxide—a green pigment that produces a well-defined anodic signal in acidic electrolytes near to +1.0 V, as can be seen in Fig. 2.15. This corresponds to the electrochemical process [147, 149]:

$$Cr_2O_{3s} + 4H_2O \rightarrow Cr_2O_7{}^{2-}{}_{aq} + 8H^+{}_{aq} + 6e^- \tag{2.13}$$

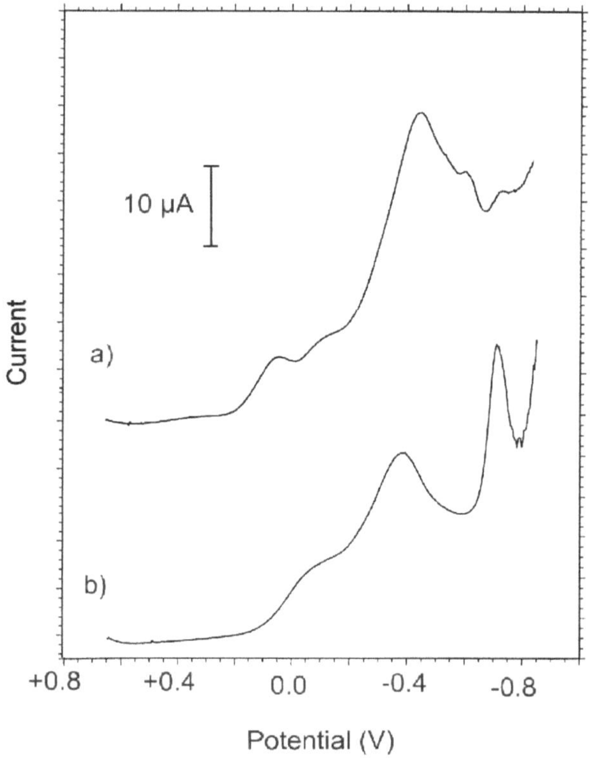

Fig. 2.14 SQWVs of PIGEs modified with: (**a**) *Caput mortum*, (**b**) Venetian *red*, immersed into 0.10 M HCl. Potential scan initiated at +0.65 V in the negative direction; potential step increment 4 mV; square wave amplitude 25 mV; frequency 5 Hz [139]

2.7 Identification of Species Via Solid-State Transformations

The electrochemical processes involving Prussian blue and organic dyes studied above can be taken as examples of solid state redox processes involving transformation of a one solid compound into another one. This kind of electrochemical reactions are able to be used for discerning between closely related organic dyes. As previously described, the electrochemistry of solids that are in contact with aqueous electrolytes involves proton exchange between the solid and the electrolyte, so that the electrochemical reaction must in principle be confined to a narrow layer in the external surface of the solid particles. Eventually, however, partial oxidative or reductive dissolution processes can produce other species in solution able to react with the dye.

The characterization of individual dyes is complicated by: (a) the existence of a significant structural similarity, and consequently, reactivity, for different dyes; (b) several dyes are composed of mixtures of pigmenting compounds, in turn having

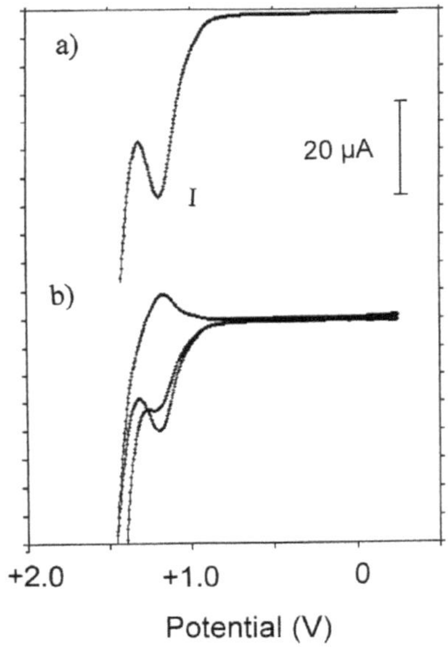

Fig. 2.15 SQWV for Cr_2O_3 immersed into 0.10 M HCl shown: (**a**) the net difference current between forward and backward pulses, and, (**b**) such currents and their sum. Potential scan initiated at +0.15 V in the positive direction. Potential step increment 4 mV; square wave amplitude 25 mV; frequency 5 Hz. Adapted from [149]

a close structural similarity. Scheme 2.1 shows structure diagrams for typical anthraquinonic and flavonoid compounds.

Figure 2.16 compares the SQWV responses obtained for silk fibbers pigmented with: (a) *granado*; (b) *alazor*; (c) curcuma, (d) weld, and (e) cochineal red in contact with acetic/acetate buffer. This voltammetry yields clearly different responses for the involved dyes, thus allowing for an electrochemical identification of pigments in samples.

The electrochemistry of anthraquinonic compounds in solution is dominated by the proton-assisted reduction of the quinone group to diphenol. Flavonoid and anthraquinonic compounds undergo electrochemical oxidations through [150–154] the orthophenol moiety, while their reduction can occur through the keto group of the benzopyrone moiety yielding hydroxyl derivatives at potentials ca. −1 V [150–154]. It should be noted, however, that the presence of additional phenol groups can complicate the electrochemical response of flavonoids. Thus, Brett and Ghica have recently reported [155] that the electrochemical oxidation of quercetin (3,3′,4′,5,7-pentahydroxyflavone) on glassy carbon electrodes occur in a cascade mechanism, related to the two catechol hydroxy groups and the other three hydroxyl groups that are all electroactive. Adsorption complicates the voltammetric response, and the final oxidation product is not electroactive and blocks the electrode surface [155]. Chalcones, further oxidized to flavonoids at potentials above +0.5 V [156–158], can also be involved.

These electrochemical processes can be accompanied by secondary reactions in solution. The presence of additional phenol groups can complicate the electrochemical response of flavonoids, giving rise to additional electrochemical processes [150–158]. This fact is of interest for analytical purposes, because

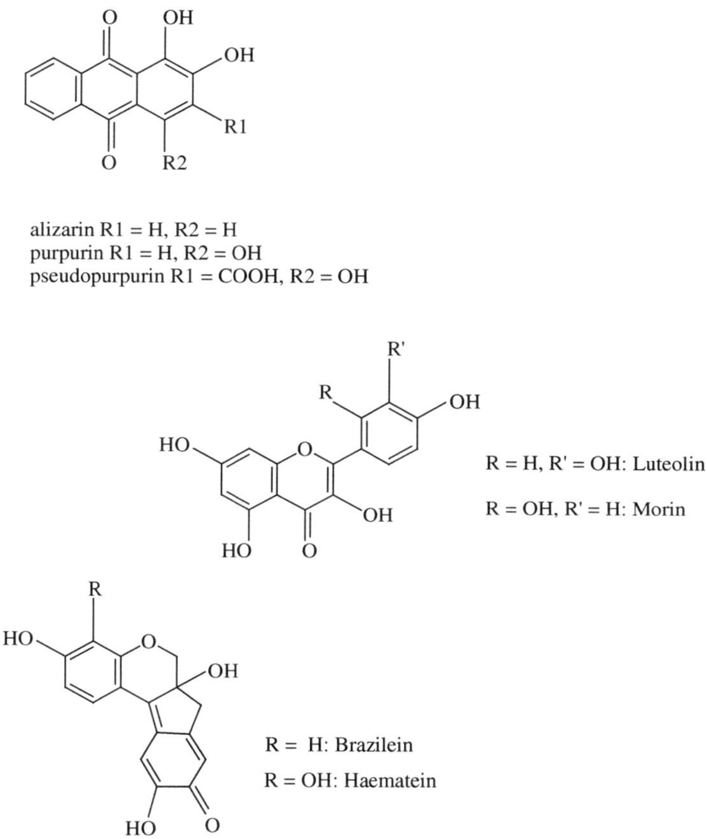

Scheme 2.1 Structural representations of typical anthraquinonic, flavonoid, and flavonoid-related dyes

additional dye-characteristic voltammetric peaks can appear. As a result, the solution phase electrochemistry of organic dyes is strongly dependent on the electrolyte composition and electrochemical conditions.

The voltammetric response of curcumin and carthamin must, in principle, be dominated by the oxidation of the phenol and/or methoxyphenol groups (see Scheme 2.2). The electrochemistry of methoxyphenols has claimed considerable attention because of their applications in organic synthesis [159–163]. As studied by Quideau et al., in aprotic media, 2-methoxyphenols are oxidized in two successive steps into cyclohexadienone derivatives [163], whereas α-(2)- and α-(4-methoxyphenoxy) alkanoic acids undergo electrochemically induced spirolactonization to develop synthetically useful orthoquinone bis- and monoketals. In the presence of methanol, the electrochemical pathway involves an initial one-electron loss, followed by proton loss, to form a monoketal radical. This undergoes a subsequent electron and proton loss coupled with the addition of alcohol to form an orthoquinone monoketal. The formal electrode potential for the second electron transfer

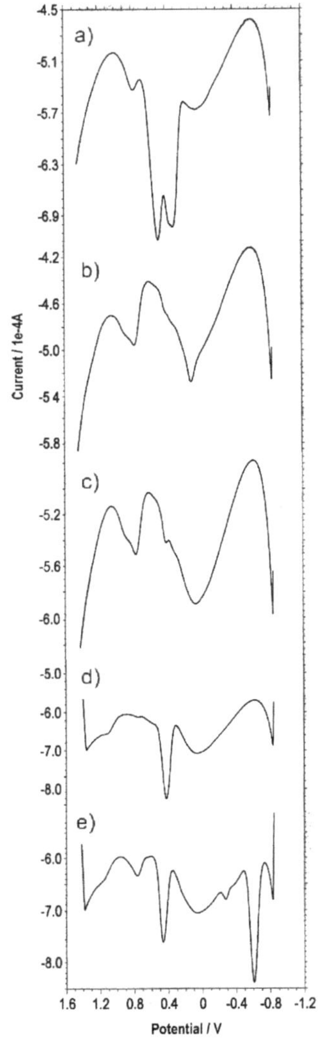

Fig. 2.16 SQWVs for silk fibers pigmented with: (**a**) *granado*; (**b**) *alazor*; (**c**) curcuma, (**d**) weld, and (**e**) cochineal *red* in contact with acetic/acetate buffer (total concentration 0.50 M) at pH 4.85. Potential step increment 4 mV; square wave amplitude 25 mV; frequency 5 Hz

is lower than that of the first electron transfer, thus resulting in an apparently unique irreversible oxidation voltammetric signal. In aqueous media, p-methoxyphenols yield p-benzoquinone.

Examples of the application of solid state electrochemistry to identifying dyes in textile samples can be provided. Thus, Fig. 2.17 compares the square wave voltammograms of: (a) saffron blank, and (b) sample from a Tibet temple, attached to fluorine-doped tin oxide (FTO) electrodes immersed into acetate buffer. After initiating the potential scan at −0.85 V in the positive direction, two separated oxidation

Scheme 2.2 Structural diagrams for curcumoid compounds curcumin and carthamin

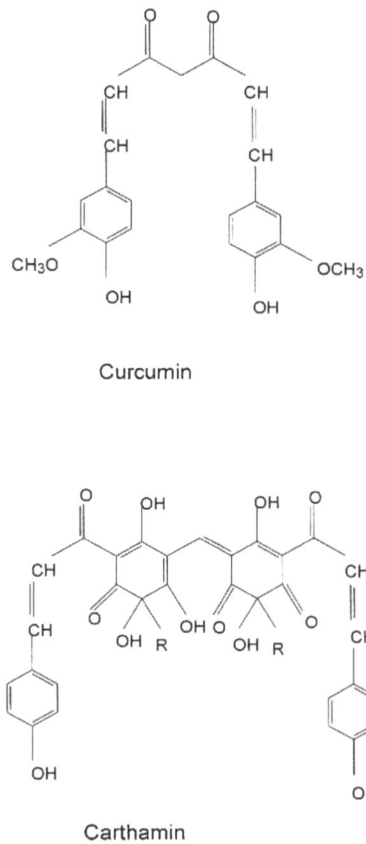

peaks at −660 and −140 mV were recorded. A close similarity was obtained between the reference saffron and the studied sample.

Application of this methodology to the analysis of materials in textiles can be illustrated by the identification of pigmenting species in brocade pieces from the church of Castellfort (Valencian Region, Spain), dated from the 25th century (see Fig. 2.18) [124]. Here, microsamples from fibers were transferred to the surface of paraffin-impregnated graphite electrodes. The voltammetric response, in contact with acetate buffer for a sample from a scapulary, is compared in Fig. 2.19 with those of carmine and Brazilwood. Examination of such voltammograms clearly suggests that the sample contains a mixture of the above dyes as pigmenting materials.

2.8 Analytical Strategies

The analytical capabilities of solid state electrochemical techniques can be increased by “chemical” and “electrochemical” procedures. By the first means, the most direct approach consists in the sequential use of different electrolytes incorporating

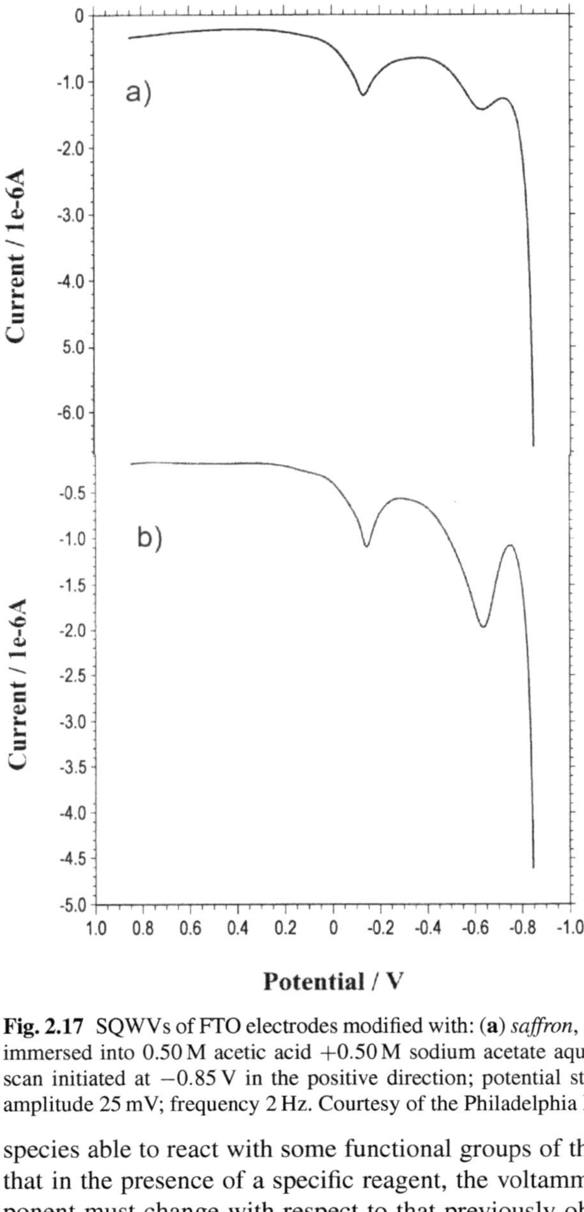

Fig. 2.17 SQWVs of FTO electrodes modified with: (**a**) *saffron*, and, (**b**) sample from Tibet temple immersed into 0.50 M acetic acid +0.50 M sodium acetate aqueous buffer at pH 4.85. Potential scan initiated at −0.85 V in the positive direction; potential step increment 4 mV; square wave amplitude 25 mV; frequency 2 Hz. Courtesy of the Philadelphia Museum of Arts

species able to react with some functional groups of the analytes. The basic idea is that in the presence of a specific reagent, the voltammetric response of each component must change with respect to that previously obtained in the absence of the reagent. By the second means, different electrochemical parameters can be varied in order to improve the electrochemical recognition of selected compounds—namely, initial and switching potentials and the potential scan rate in cyclic voltammetry, and the initial and final potentials, as well as potential step increment, square wave amplitude, and frequency in square wave voltammetry. Additionally, application of

Fig. 2.18 Image of a scapulary from the mediaval textile workshops in Morella (Valencian Region, Spain) existing in the church of Castellfort (Valencian Region, Spain), dated from the 25th century

polarization steps (repeatedly cycling the potential between two given values or applying a constant potential during a given "electrogeneration time" can result in significant differences between the pre- and post-treatment voltammetric record). Thus, the voltammetric response of anthraquinonic and flavonoid compounds is particularly sensitive to the application of negative polarization potentials—a feature previously noted [122] and also observed by Grygar et al. [123] and Dai and Shin [164] for anthraquinone-type compounds.

The first strategy can be illustrated by the effect of the addition of boric acid (H_3BO_3) or $AlCl_3$ to acetate buffers with respect to the response of flavonoid dyes [124]. We know that boric acid coordinates with compounds containing o-diphenol groups, which form stable adducts in solution [165]. In turn, Al^{3+} ions form strong complexes with flavonoids containing 3- and 5-hydroxy groups [166]. Assuming that solid compounds can also form surface-confined adducts with those coordinating agents, one can expect that the response of solid dyes containing o-diphenol and/or 3- and 5-hydroxy groups should change in the presence of the above species.

Experimental data for flavonoid dyes confirmed these expectances [124]. Thus, the voltammetric response was expected—the square wave voltammograms of flavonoids containing 3- and 5-hydroxy groups change from 0.25 M HAc + 0.25 M NaAc to that plus 0.05 M $AlCl_3$. The obtained results are illustrated in Fig. 2.20. First, the SQWVs of weld-modified PIGEs in the absence (Fig. 2.20a) and presence (Fig. 2.20b) of Al^{3+} are shown. Coordination with aluminium apparently blocks electron-transfer processes, and the peaks are significantly decreased. For logwood, the contrast between the square wave voltammograms in the absence and presence

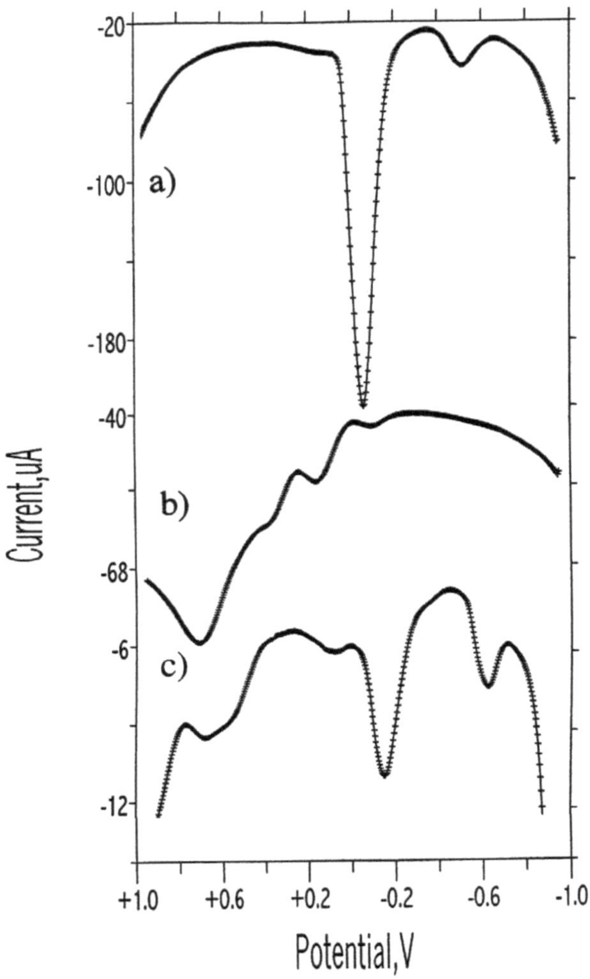

Fig. 2.19 Square wave voltammograms for: (**a**) carmine, (**b**) Brazilwood, and (**c**) a fiber sample from a mediaeval scapulary found in the church of Castellfort (Valencian Region, Spain). Electrolyte: 0.50 M sodium acetate buffer. Potential step increment 4 mV; square wave amplitude 25 mV; frequency 5 Hz. Adapted from [124]

of Al^{3+} (illustrated in Fig. 2.20c and d) is also remarkable. In this case, voltammetric peaks are all decreased except the peak at +0.03 V, which is abruptly enhanced. This probably results from the formation of an adsorbate of the aluminum complex on the electrode surface—a situation similar to that existing in alizarin-type aluminium complexes [124]. A similar situation occurs when boric acid is added to the electrolyte with regard to dyes containing o-phenol moieties, due to the strong complexation of such groups by borate units. Structural diagrams for the aforementioned coordination are depicted in Scheme 2.3.

Fig. 2.20 SQWVs (first scan) of weld (**a**, **b**) and logwood (**c**, **d**) immersed into 0.25 M HAc + 0.25 M NaAc (**a**, **c**) and that electrolyte plus 0.05 M $AlCl_3$ (**b**, **d**). Potential step increment 4 mV; square wave amplitude 25 mV; frequency 15 Hz [124]

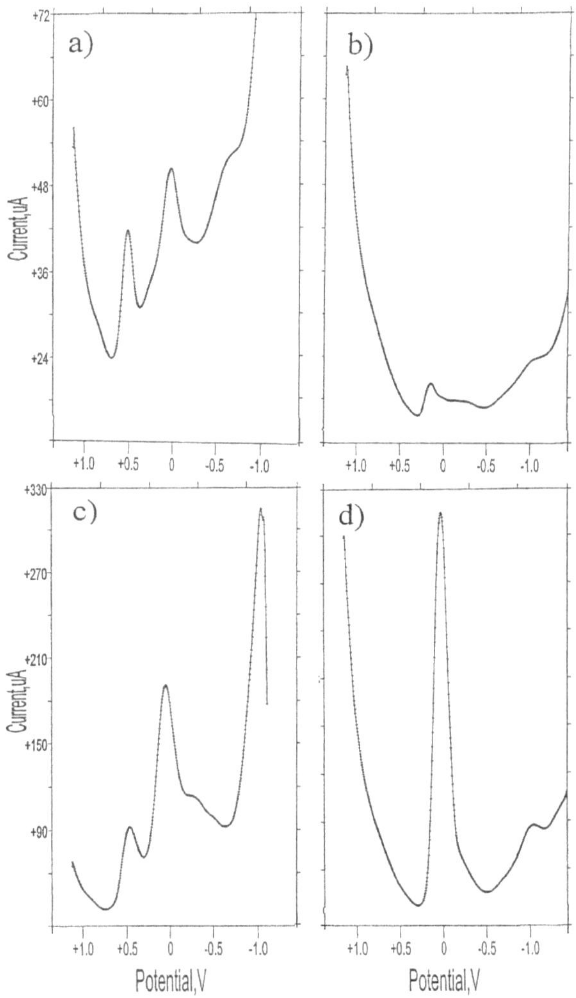

The described procedures are able to distinguish organic dyes in solid samples. However, in several cases the distinction between some pigments can be made uncertain by matrix effects. These can be attributed mainly to (a) the presence of additional peaks associated with other dyes and/or priming and protective organic compounds coexisting in the sample; (b) occurrence of cross reactions between the intermediate products of electron-transfer processes; (c) modification of the electrochemical response of the dyes as an effect of complexation reactions with metal ions from inorganic pigments; or (d) appearance of unexpected peaks associated with alteration products.

The application of constant potential steps results in the generation of new solid compounds whose electrochemical response should differ from the original

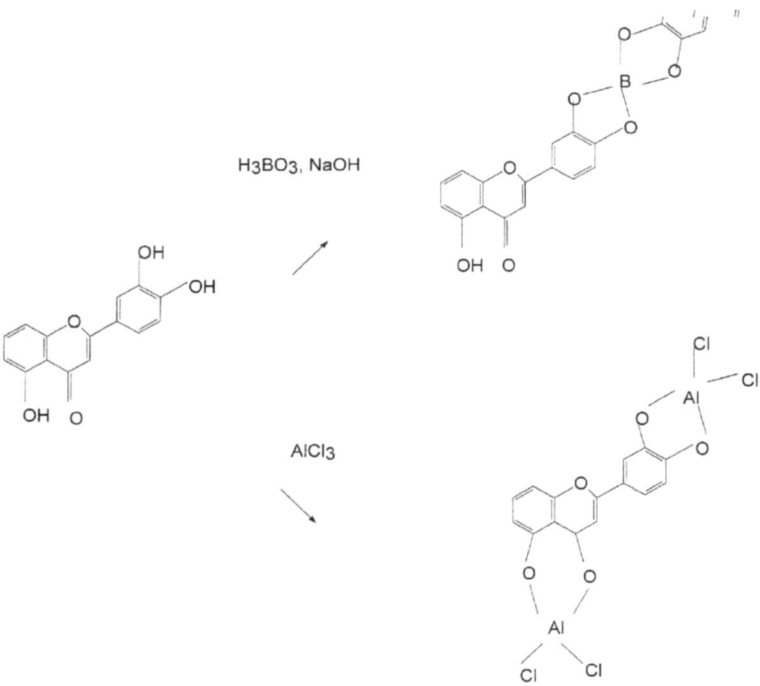

Scheme 2.3 Adduct formation between flavonoids and boric acid and aluminum chloride [124]

material. The use of the responses of the parent material and its electrochemically generated derivative can be used for discerning similar materials.

This strategy is illustrated in Figs. 2.21 and 2.22 for the case of alizarin. The parent dye exhibits two main peaks at +0.40 and −0.60 V, corresponding respectively to the oxidation of the o-phenol moiety and the reduction of the quinone group. After application of a reduction step at −1.50 V, the resulting polyhydroxy species produces overlapping oxidation signals at +0.5 and −0.07 V.

Oxidative treatments also yield significant changes in the voltammetric response. As shown in Fig. 2.22, the response of alizarin changes remarkably after application of an anodic step. Oxidation of alizarin should make polyquinone species able to be electrochemically reduced at potentials differing from those at which the parent alizarin reduces. As a result, the voltammetric pattern also changes drastically. Application of this strategy increases the number of parameters that can be used as diagnostic criteria for identifying dyes.

It should be noted, however, that the changes in the voltammetric response are conditioned by the nature of the extent of the redox reaction across the solid. Thus, for organic solids in contact with aqueous electrolytes, and using the aforementioned model of Lovric, Scholz, Oldham, and co-workers [115–118], the propagation of the redox reaction should involve proton hopping coupled with electron hopping between adjacent immobile molecules [119–125]. Chronoamperometric data

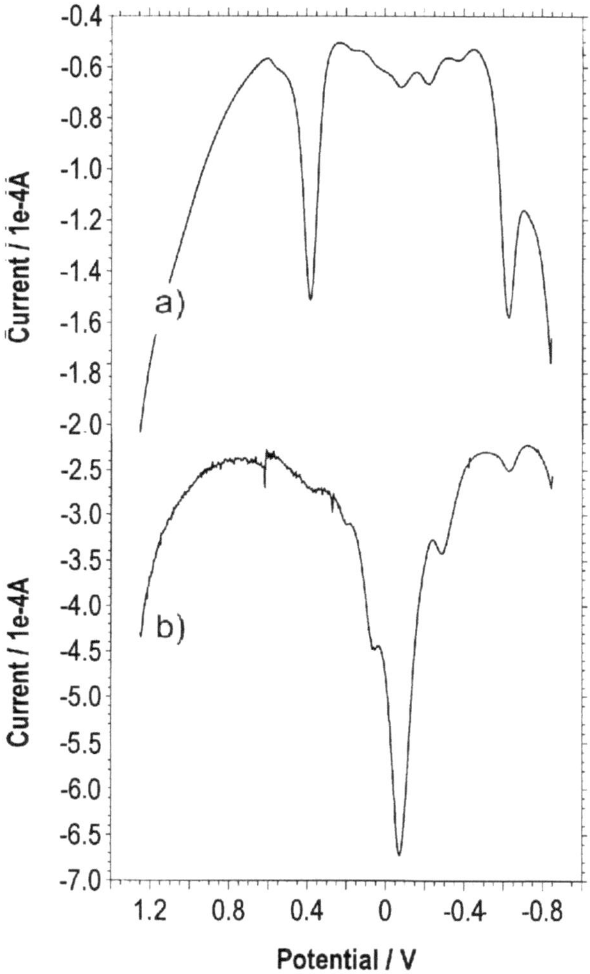

Fig. 2.21 Square wave voltammograms for alizarin-modified paraffin-impregnated graphite electrodes immersed into 0.50 M potassium phosphate buffer (pH 7.0): (**a**) before, and (**b**) after application of a potential step of −1.50 V during 15 min. Potential initiated at −0.85 V in the positive direction. Potential step increment 4 mV; square wave amplitude 25 mV; frequency 5 Hz

reveals that electron diffusion in organic compounds in contact with aqueous buffers is significantly faster than proton diffusion [167]. Accordingly, application of a reductive step could produce a situation similar to that ideally represented in Fig. 2.23. The inverse case, when electron transport was clearly faster than proton transport, is schematized in Fig. 2.24. The voltammetric response of these systems should be different to that of the parent solid.

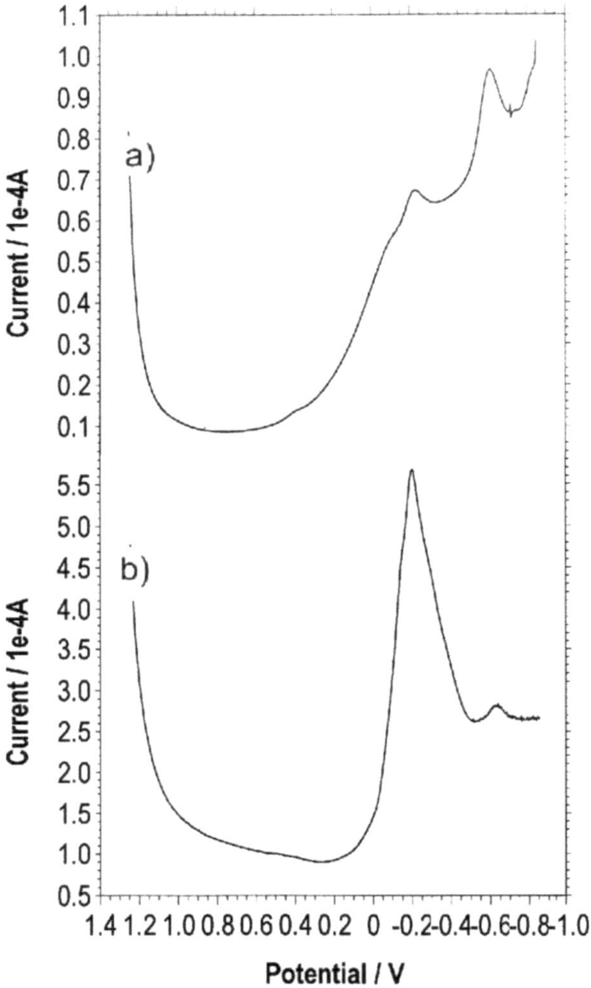

Fig. 2.22 Square wave voltammograms for alizarin-modified paraffin-impregnated graphite electrodes immersed into 0.50 M potassium phosphate buffer (pH 7.0): (**a**) before, and (**b**) after application of a potential step of +1.50 V during 15 min. Potential initiated at +1.25 V in the negative direction. Potential step increment 4 mV; square wave amplitude 25 mV; frequency 5 Hz

Topographic AFM examination of organic dyes under the application of reductive and oxidative potentials appears to be consistent with the above modeling [168]. This can be seen in Fig. 2.25 where in situ AFM images from the upper face and sides of crystals of indigo (a) before, and (b) after application of a linear potential step between 0.0 and +0.75 V at a potential scan rate of 10 mV/s are shown. Here,

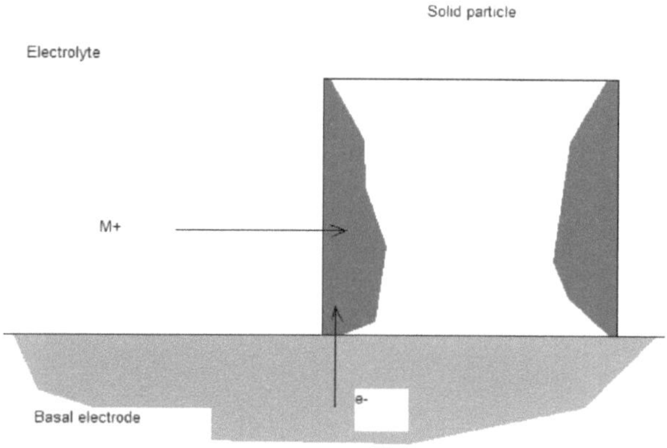

Fig. 2.23 Schematics representation of the electroactive layer in the case of the electrochemical reduction of a solid when ion diffusion through the solid is allowed—electron diffusion is highly hindered

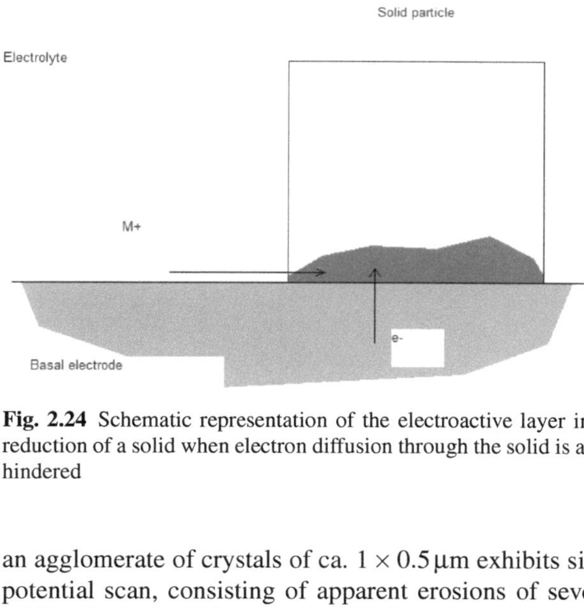

Fig. 2.24 Schematic representation of the electroactive layer in the case of the electrochemical reduction of a solid when electron diffusion through the solid is allowed and ion diffusion is highly hindered

an agglomerate of crystals of ca. $1 \times 0.5\,\mu\text{m}$ exhibits significant changes during the potential scan, consisting of apparent erosions of several regions of lateral sides, while other regions become apparently extended. A possible scheme for explaining these features may consist of an initially topotactic process, eventually accompanied by exfoliation and restacking, and followed by crystal restructuring and eventually phase transition. The maximum extent of the lateral distortions of crystals can be estimated as long as ca. 50–80 nm—a value that is in close agreement with that calculated for the breadth of the diffusion layer from chronoamperometric data, taking $D = D_H = 2 \times 10^{-10}\,\text{cm}^2/\text{s}$. [167, 168].

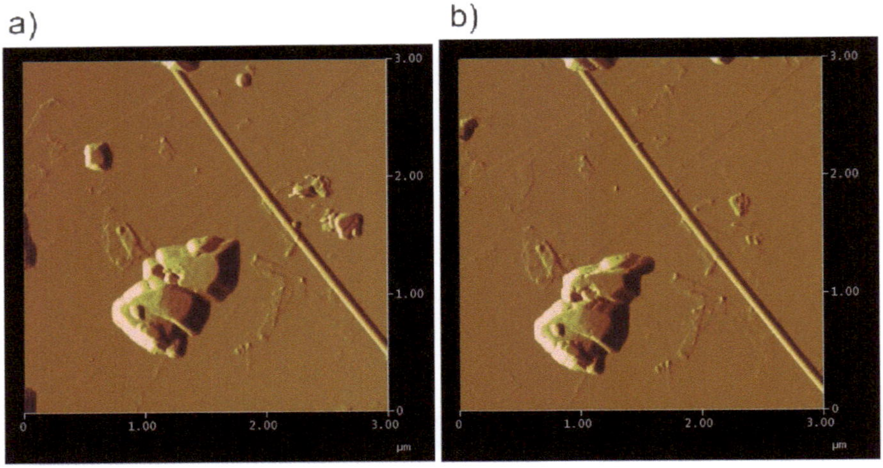

Fig. 2.25 AFM of indigo crystals adhered to a graphite plate in contact with 0.50 M sodium aqueous acetate buffer (**a**, **c**) before, and (**b**, **d**) after application of a linear potential scan between 0.00 and +0.75 V at a sweep rate of 10 mV/s [168]

Chapter 3
Resolution of Multicomponent Systems and Speciation

3.1 Introduction

In the previous chapter, methodologies for identifying individual species in works of art have been presented for cases where only one (or one main) electroactive component is present. Such electrochemical methods, however, can be extended to cases in which several electroactive species contribute to the observed electrochemical response. Resolution of multicomponent systems, considered here as a possibility for simultaneously identifying different electroactive compounds in the sample, can be achieved by applying different methodologies. These can, in principle, be divided into: (a) "electrochemical" methodologies, based on the variation of electrochemical parameters and handling of electrochemical data, and, (b) "chemical" methodologies, based on the selective choice of chemical inputs (pH variation, complexing electrolytes) for improving the electrochemical discrimination of the components of the sample. Within "electrochemical" methodologies, we can distinguish: (i) the use of pattern recognition criteria by taking shape-dependent parameters for individual electrochemical signals; (ii) application of bi-parametric classification schemes; and (iii) the application of multivariate chemometric methods. This chapter will be devoted to describing these approaches in order to identify different components in samples from works of art and archaeological artifacts. Obviously, electrochemical methods can be complemented with nonelectrochemical techniques to obtain analytical information demanded for conservators and restorers [169–172].

It is important to note that the application of electrochemical methods to the analysis of samples of art objects and archaeological artifacts allows much more than only simple identification of certain constituents: advanced methods of speciation may provide information about constituents that are only slightly differing in their composition, or for which there are only slight differences in the matrices in which the components are embedded. Further, redox speciation—and in the case of solid samples, phase speciation—can be used to derive information on production processes or corrosion (deterioration) of the components in the time that passed since their formation. The second part of this chapter is devoted to illustrating the capabilities of advanced speciation strategies.

A. Doménech-Carbó et al., *Electrochemical Methods in Archaeometry, Conservation and Restoration*, Monographs in Electrochemistry,
DOI 10.1007/978-3-540-92868-3_3, © Springer-Verlag Berlin Heidelberg 2009

3.2 Analysis of Single Multicomponent Systems

In several cases, samples from works of art contain different electroactive species that provide separated signals. This is frequent in paint samples where different pigments—and eventually, alteration products—exist. In general, such components occur as microparticles highly diluted in a nonelectroactive matrix so that independent voltammetric responses are, in principle, recorded. This is the case for a sample of medieval glaze from Manises (Valencia, Spain), the square wave voltammogram of which is depicted in Fig. 3.1. Here, cathodic peaks at +0.70, −0.08, −0.45, and −0.56 V appear. The first two peaks correspond, respectively, to the reduction of Mn(IV) and Sn(IV), while the third and fourth peaks can be ascribed to the reduction of Sn(II) and Pb(II) to the corresponding metals [173]. This response provides information on the composition of the glaze and denotes the presence of manganese as a coloring agent in that sample [128].

Figure 3.2 shows the square wave voltammetric response for: (a) azurite, (b) tenorite (CuO), and (c) a sample from the frescoes in the ceiling vault of the *Sant Joan del Mercat* church in Valencia (Spain)—all attached to PIGEs and immersed into 0.50 M acetate buffer. These frescoes, painted by Antonio Acisclo Palomino in 1707 suffered considerable damage by fire during the Spanish civil war in 1936. As can be seen in Fig. 3.3, a substantial part of the paintings was entirely destroyed, while most of the surviving areas presented significant chromatic changes [133, 172, 174].

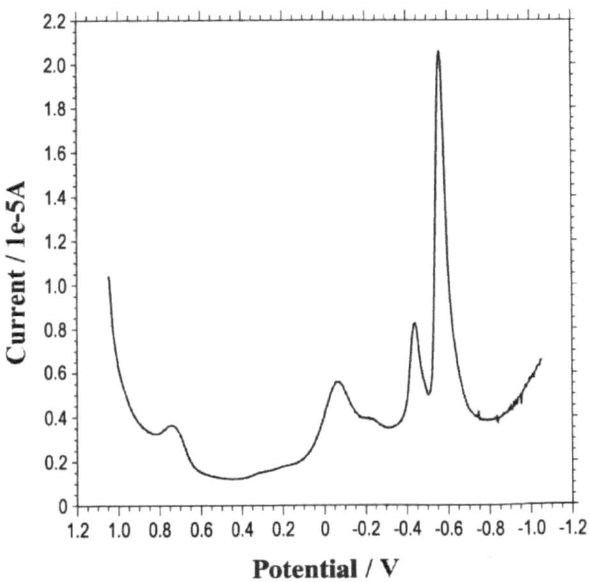

Fig. 3.1 Square wave voltammogram for a sample of mediaeval glaze from Manises (Valencia, Spain). Electrolyte 0.50 M sodium acetate buffer, pH 4.85. Potential step increment 4 mV; square wave amplitude 25 mV; frequency 5 Hz

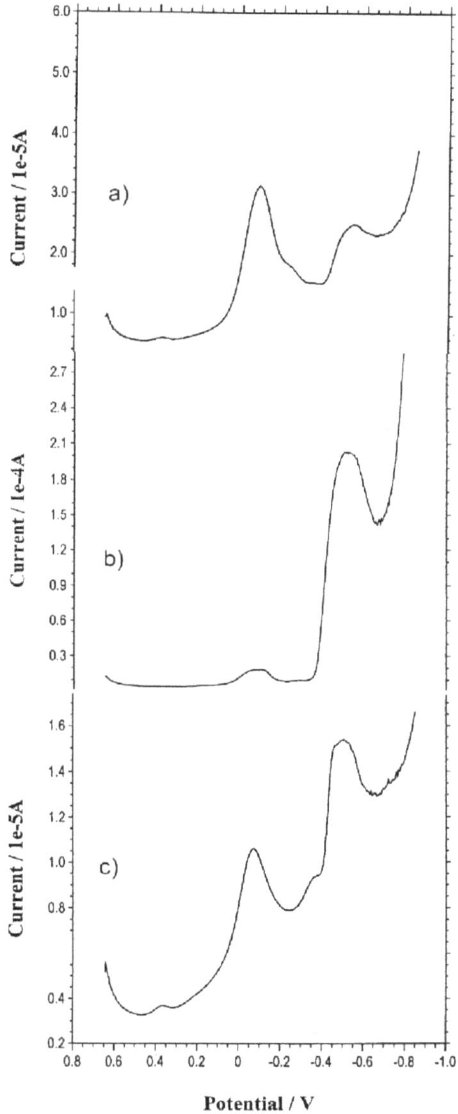

Fig. 3.2 Square wave voltammograms of: (**a**) azurite, (**b**) CuO, and (**c**) a sample from the *Sant Joan del Mercat* church in Valencia (Spain), attached to PIGEs and immersed in 0.50 M sodium acetate buffer (pH 4.85). Potential scan initiated at +0.65 V in the negative direction. Potential step increment 4 mV; square wave amplitude 25 mV; frequency 5 Hz [133]

SQWVs of "green" paint layers confirmed the presence of azurite and azurite plus smalt mixtures accompanied by tenorite (CuO) in several samples as a result of thermal decomposition of azurite. Since azurite decomposition starts at 345°C with a loss of CO_2 and water, slowly yielding CuO which is then converted to cuprite (Cu_2O) at 840°C [175], and since no traces of this last mineral were detected in any sample, one can conclude that the effective temperature reached by the pigments in such samples ranged from 350 to 840°C.

Fig. 3.3 Image of a portion (area ca. 1 m^2) of the damaged Palomino's frescoes in the vault of the *Sant Joan del Mercat* church in Valencia (Spain) [133]

3.3 Criteria for Pattern Recognition

In most cases, electrochemical data consists of a series of well-separated peaks due to the reduction/oxidation of different electroactive species existing in the sample. Identification of the component responsible for each peak can, in principle, be derived from the peak potential, in a given electrolyte, under fixed electrochemical conditions. This parameter, however, depends (in the case of solid materials) on the shape and size distribution (i.e., in a wide meaning, its granulometric characteristics) of the particles of electroactive compound. Apart from this, matrix effects (complexing ions existing in the sample) can also modify the observed voltammetric response, which, in turn, is also frequently shifted by an uncompensated ohmic drop in the cell and capacitive effects. As a result, it is desirable to use additional parameters defining diagnostic criteria for shape recognition. On the basis of criteria used by Sybrandt and Perone [176], different electrochemical parameters can be used for pattern recognition in cyclic voltammetric curves. Among others, one can mention:

(a) Onset potentials defined by prolonging the almost-linear regions in the ascendant (E_{onset}) and descending (E'_{onset}) branches of the reduction (or oxidation) peak to intersect the potential axis. It should be noted, however, that these onset potentials have no strictly defined meaning.
(b) Half-peak width $W_{1/2}$, defined as the potential separation between the points of the peak whose current is just one half of the peak current.
(c) The semi-half-peak widths, defined as the differences between the peak potential and the potentials of the above points in the increasing $(\Delta E_{p/2})_1$ and decreasing $(\Delta E_{p/2})_2$ portions of the voltammetric peak. The quotient $r = (\Delta E_{p/2})_1/(\Delta E_{p/2})_2$ between these two parameters defines a symmetry factor for the voltammetric peak.

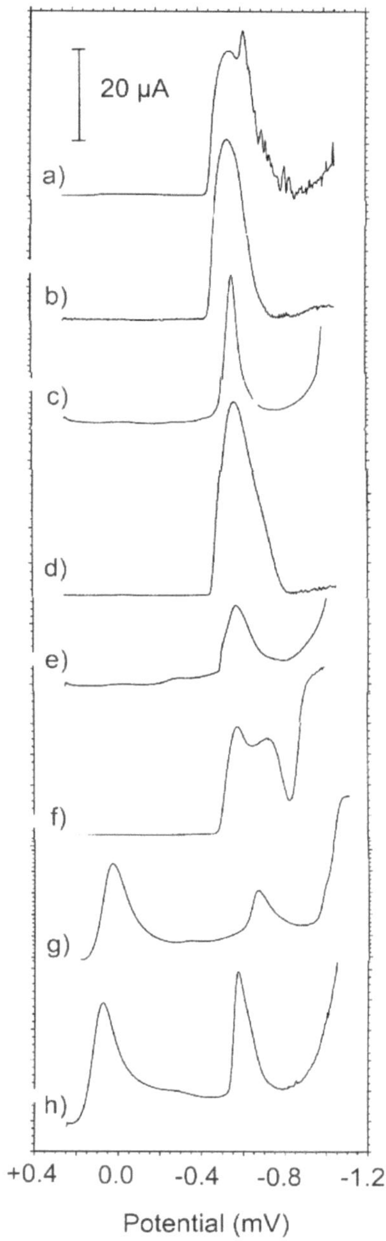

Fig. 3.4 Square wave voltammograms of microsamples of: (**a**) litharge, (**b**) massicot, (**c**) *Naples yellow*, (**d**) *lead white*, (**e**) *tin-lead yellow*, (**f**) minium, (**g**) *chrome orange*, and (**h**) *chrome yellow* attached to parafin-impregnated graphite electrodes. Electrolyte: 0.50 M sodium acetate buffer, pH 4.85. Potential step increment 4 mV; square wave amplitude 25 mV; frequency 15 Hz [177]

Table 3.1 Diagnostic criteria for characterizing lead pigments via voltammetry of microparticles using deposits of the pristine pigments on paraffin-impregnated graphite electrodes. Data from square-wave voltammograms at a potential step increment of 4 mV, square-wave amplitude of 25 mV, and frequency of 15 Hz. All potentials refer to AgCl (3M NaCl)/Ag. Electrolyte, 0.50 M acetate buffer, pH 4.85

Pigment	E_p (mV)	$W_{1/2}$ (mV)	$(\Delta E_{p/2})_1$ (mV)	$(\Delta E_{p/2})_2$ (mV)	r
Chrome Orange	−605	115	35	80	0.44
Chrome Yellow	−570	85	25	60	0.42
Lead White	−555	180	65	115	0.56
Litharge	−540	160	65	95	0.68
Massicot	−540	145	60	85	0.70
Minium	−545	155	50	105	0.48
Naples Yellow	−540	65	30	35	0.86
Tin-lead Yellow	−570	135	60	75	0.80

(d) Separation between the potentials for which current is 3/4 and 1/4 of the peak current in the main reduction wave, $E_{3/4} - E_{1/4}$. This parameter, representing the "mean" inclination of the current/potential curve, is of interest for characterizing voltammetric processes displaying highly overlapping peaks [176].

The convenience of using multiparametric pattern recognition criteria can be illustrated by the problem of identifying lead pigments in pictorial samples [126]. Figure 3.4 compares square-wave voltammograms for: (a) litharge, (b) massicot, (c) Naples yellow, (d) lead white, (e) tin-lead yellow, (f) minium, (g) chrome orange, and (h) chrome yellow attached to paraffin-impregnated graphite electrodes immersed in acetate buffer. Lead pigments exhibit a prominent reduction peak at about −0.55 V corresponding to the reduction of the corresponding lead compound into lead metal. For instance, for litharge or massicot,

$$PbO_s + 2H^+{}_{aq} + 2e^- \rightarrow Pb_s + H_2O \tag{3.1}$$

Only chrome yellow and chrome orange (lead chromates) exhibit an additional peak at +0.45 V, attributable to the reduction of chromate ions:

$$PbCrO_{4s} + 8H^+{}_{aq} + 3e^- \rightarrow Cr^{3+}{}_{aq} + Pb^{2+}{}_{aq} + 4H_2O \tag{3.2}$$

As can be seen in Fig. 3.4, peak potentials for lead-centered reduction processes are similar, so pigment identification can be improved by considering a series of parameters, as summarized in Table 3.1.

3.4 Bi-Parametric Data Analysis

Combination of selected pairs of variables in two-dimensional (2D) diagrams enables a separation between different electroactive materials. This can be seen in Fig. 3.5, where a representation of $(E_{3/4} - E_{1/4})$ vs. $(E'_{onset} - E_{onset})$ measured in

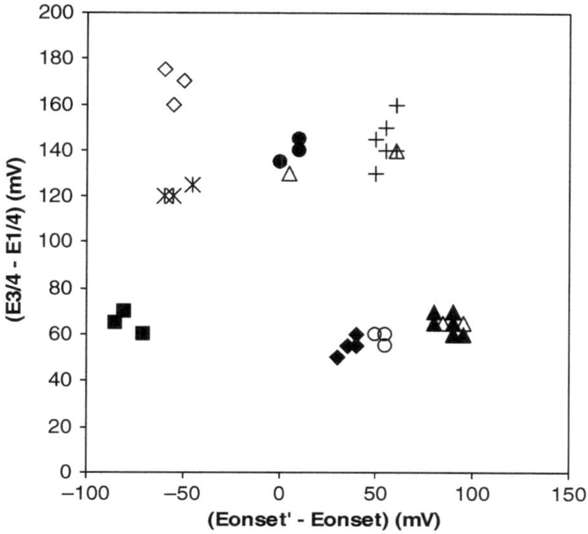

Fig. 3.5 Representation of ($E_{3/4} - E_{1/4}$) vs. ($E'_{onset} - E_{onset}$) for lead pigments, model paint specimens, and real samples. Data from CVs recorded in the AFM cell for nanosamples attached to a graphite plate. Electrolyte: 0.50 M acetate buffer. Potential scan rate 100 mV/s. *Solid rhombs* = chrome orange; *solid squares* = chrome yellow; *solid triangles* = lead white; *rohmbs* = litharge; *stars* = massicot; *solid circles* = minium; *crosses* = Naples yellow; *circles* = tin-lead yellow. Real samples for different paint specimens (see details in ref. [126]) are marked by *triangles*

cyclic voltammograms for lead pigments, model paint specimens, and real samples attached to a graphite plate in an electrochemical atomic force microscopy cell immersed in acetate buffer is depicted [126]. Here, the inherent capabilities of electrochemical AFM are used for recording the voltammetric response of nanosamples of pictorial specimens.

Although the presence of binding media and other materials accompanying pigment in samples influences voltammetric signals, under appropriate selection of the electrochemical parameters and operating conditions [126], voltammetric responses become pigment-characteristic and fall in clearly separated regions of the 2D diagram.

In favorable cases, detailed analysis of electrochemical parameters can provide information on the composition of complex systems such as pigments plus binding media in paint samples. In general, voltammetric profiles of pigments can be modified to any extent by the presence of other materials in the sample—binding media in particular. As it is well-known, pigments are accompanied in such samples by bindings and other materials—namely, proteinaceous materials and drying oils, either alone or mixed together.

This can be seen in Fig. 3.6, where square-wave voltammograms for synthetic pictorial specimens contain recommended dosages [177] of lead white, plus: (a) casein and (b) sunflower oil. As depicted in Fig. 3.3, the response of the pristine pigment consisted of a unique peak at −0.56 V. The pictorial specimens

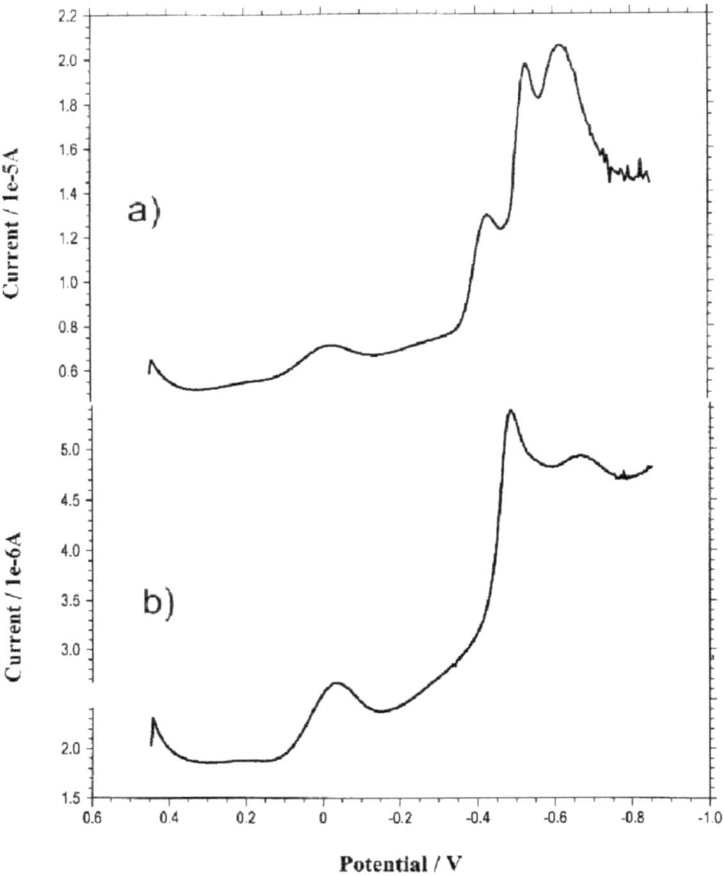

Fig. 3.6 SQWVs for synthetic specimens containing lead white plus (**a**) casein and (**b**) sunflower oil, immersed into 0.50 M sodium acetate buffer, pH 4.85. Potential step increment 4 mV; square wave amplitude 25 mV; frequency 15 Hz [177]

provided significantly binder-dependent profiles. Thus, casein peaks at −0.43, −0.54, and −0.62V appear, while square-wave voltammograms provided a sharp peak at −0.48 V for oil-containing specimens followed by a shoulder at ca. −0.65 V.

The above voltammetric responses can be rationalized by assuming that the response of the grains of pristine pigments is superimposed to that of the ionomeric layer containing lead-binder compounds [178, 179]. For lead carbonate, this can be represented as [126, 177]

$$PbCO_{3s} + 2H^{+}{}_{aq} + \{binder\} \rightarrow \{Pb^{2+} - binder\} + H_2O + CO_2 \qquad (3.3)$$

$$\{Pb^{2+} - binder\} + 2H^{+}{}_{aq} + 2xe^{-} \rightarrow xPb_s + (1-x)Pb^{2+}{}_{aq} + \{2H^{+} - binder\} \qquad (3.4)$$

$$Pb^{2+}{}_{aq} + 2e^- \rightarrow Pb_s \tag{3.5}$$

In these equations, {} represents species confined in the ionomeric layer surrounding the grains of pigment. In the case of proteinaceous binders, Pb^{2+} ions are probably coordinated to any extent to donor sites of the binder, so that Eq. 3.3 should be replaced by

$$Pb^{2+} - \{binder\} + 2e^- \rightarrow Pb_s + \{binder\} \tag{3.6}$$

Additionally, the ionomeric layer can form a barrier for charge diffusion surrounding the particles of pigment. As a result, the rate of the overall electrochemical process described by Eqs. 3.3, 3.4 and 3.5 is modified. Our data suggest that the peak at ca. −0.45 V is attributable to the reduction of free (Eq. 3.5) or binder-coordinated (Eq. 3.6) Pb^{2+} ions, while peaks at ca. −0.65 V must correspond to the reduction scheme described by Eqs. 3.3 and 3.4.

Figure 3.7 shows a 2D diagram corresponding to the plot of $(E_{onset} - E_p)$ vs. $W_{1/2}$ for the peak at −0.45 V for lead white plus proteinaceous binders (squares), lead white plus oils (triangles), minium plus proteinaceous binders (solid squares), and Naples yellow plus oils (solid triangles). In this diagram, one can see that data points for pigments plus proteinaceous binder samples fall in a region separated from that where data points for pigment plus oil binders are grouped.

Characterization of the binding media can be improved using plots of the potential separation between the peaks at −0.45 and −0.65 V, ΔE_p, vs. $W_{1/2}$ for a peak at −0.45 V. As shown in Fig. 3.8, data points for two proteinaceous binders, casein and bovine gelatin, applied to different pigments (lead white, minium, Naples yellow)

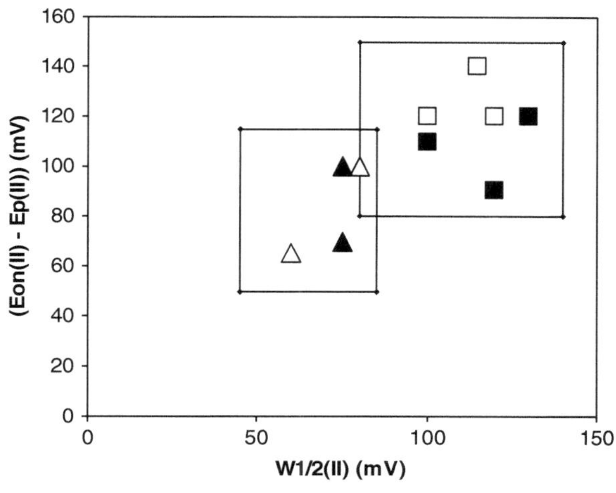

Fig. 3.7 Plot of $(E_{onset} - E_p)$ vs. $W_{1/2}$ for the peak at −0.45 V for different lead pigment plus binder specimens: lead white plus proteinaceous binders (*squares*), minium plus proteinaceous binders (*solid squares*), lead white plus oils (*triangles*), and Naples yellow plus oils (*solid triangles*). From square-wave voltammetric data using the conditions described in the caption of Fig. 3.3

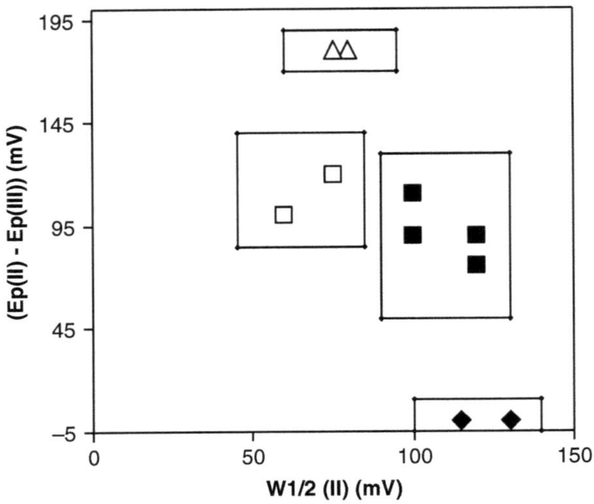

Fig. 3.8 Plot of the potential separation between the peaks at −0.45 and −0.65 V, ΔE_p vs. $W_{1/2}$ for peak at −0.45 V for different lead pigment plus binder specimens: *solid squares* = casein or bovine gelatin plus lead white or minium = *squares*: lead white and Naples yellow plus poppy oil; *triangles* = sunflower plus lead white and Naples yellow; *solid rhombs* = egg plus lead white and minium. From square-wave voltammetric data using the conditions from Fig. 3.3

fall in a narrow central region of the diagram. Data points for sunflower oil, poppy oil, and egg are in turn grouped in other, well-separated regions of the diagram.

In several instances, data analysis is complicated by the possible coexistence of different electroactive products in the sample having a similar electrochemical response. A typical case is provided by copper pigments and their alteration products in paint layers or metallic artifacts submitted to deterioration processes. Thus, Fig. 3.9a–d shows the square wave voltammograms for two typical pigments—azurite (a) and verdigris (c)—and two typical products of alteration of copper and bronze artifacts—cuprite (b), and atacamite (d)—all in contact with phosphate buffer. This last one is one of the crystalline forms of copper trihydroxychloride, which is a product of the alteration of pigments and copper pieces and eventually can also be used as a pigment [180].

In all these cases, a main reduction peak appears ca. −0.10 V so that a direct discrimination between the different species is uneasy. The situation in real samples is even more complicated because, as reviewed by Scott [181] and Tennenth [182], the alteration of bronze and copper artifacts produces (depending on the weathering conditions) different products, mainly cuprite and different minerals of the atacamite group (atacamite, paratacamite, clinoatacamite, botallakrite), accompanied by nantokite and ocasionally other rare minerals such as calumetite $(Cu(OH, Cl)_2 \cdot 2H_2O)$, anthonyte $(Cu(OH, Cl)_2 \cdot 3H_2O)$, and connellite $(Cu_{19}(SO_4)Cl_{14}(OH)_{32} \cdot 3H_2O)$ [181]. Another frequent product of alteration in bronze disease, nantokite (CuCl), exhibits a clearly discernable response consisting of a reduction peak at −0.36 V as shown in Fig. 3.10.

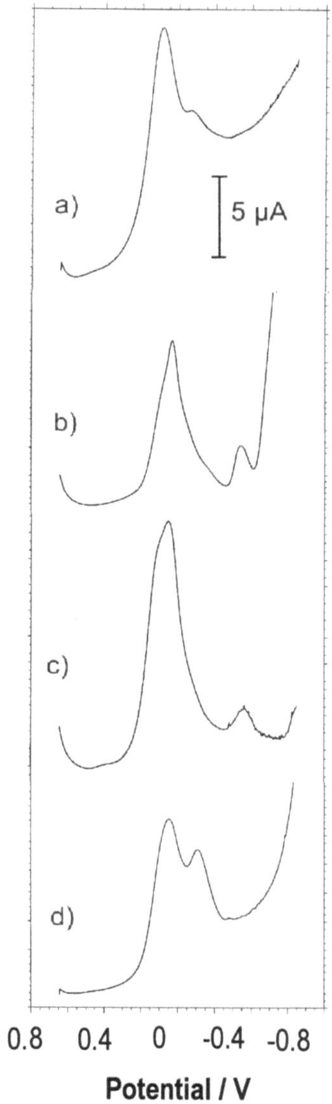

Fig. 3.9 SQWVs for (**a**) azurite, (**b**) cuprite, (**c**) verdigris, and (**d**) atacamite, in contact with 0.50 M potassium phosphate buffer, pH 7.4. Potential scan initiated at +0.45 mV in the negative direction. Potential step increment 4 mV; square wave amplitude 15 mV; frequency 2 Hz [133]

For solving mixtures of copper products, parameters derived from Tafel analysis of voltammetric curves were used [127, 133, 183]. The use of this formalism requires: (i) that electrochemical processes for individual analytes can be taken as independent, and (ii) strong overlapping of voltammetric curves. Both conditions apply in our studied samples because these systems can be taken as constituted by separated micro- or nanograins of different solid compounds. Tafel analysis is based on the assumption that the rising portion of voltammetric curves can, in general, be

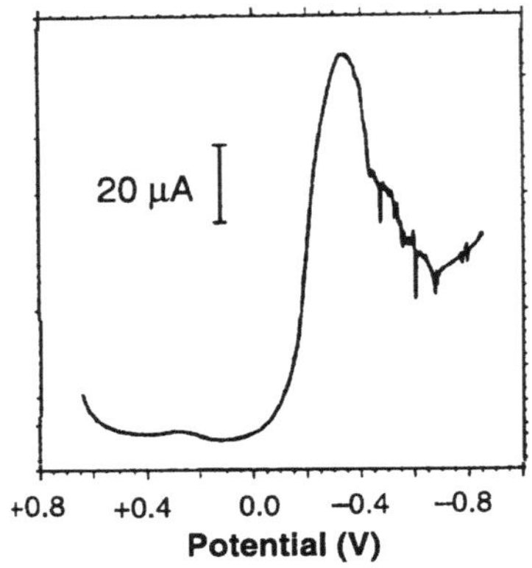

Fig. 3.10 Square-wave voltammogram of nantokite (CuCl) attached to PIGE in contact with 0.50 M potassium phosphate buffer, pH 7.0. Potential scan initiated at +650 mV in the negative direction. Potential step increment 4 mV; square wave amplitude 25 mV; frequency 5 Hz. Adapted from [183]

approached by an exponential dependence of the current on the applied potential regardless of the electrochemical mechanism. This is strictly valid for irreversible charge transfer processes involving species in solution phases, for which the current at the foot of the voltammetric peak satisfies [79, 80]

$$i = Gn(\alpha n_a)^{1/2} ckD \exp\left[-\frac{\alpha n_a F(E - E_i)}{RT}\right] \tag{3.7}$$

In this equation, αn_a represents the product of the coefficient of electron transfer (α) by the number of electrons (n_a) involved in the rate-determining step, n the total number of electrons involved in the electrochemical reaction, k the heterogeneous electrochemical rate constant at the zero potential, D the coefficient of diffusion of the electroactive species, and c the concentration of the same in the bulk of the solution. The initial potential is E_i and G represents a numerical constant. This equation predicts a linear variation of the logarithm of the current, $\ln i$, on the applied potential, E, which can easily be compared with experimental current-potential curves in linear potential scan and cyclic voltammetries. This type of dependence between current and potential does not apply to electron transfer processes with coupled chemical reactions [186]. In several cases, however, linear $\ln i$ vs. E plots can be approached in the rising portion of voltammetric curves for the solid-state electron transfer processes involving species immobilized on the electrode surface [131, 187–191], reductive/oxidative dissolution of metallic deposits [79], and reductive/oxidative dissolution of insulating compounds [147, 148]. Thus, linear potential scan voltammograms for surface-confined electroactive species verify [79]

$$i = \frac{n^2F^2 vA\Gamma(b_{ox}/b_{rd}) \exp\left[-\frac{nF(E-E^{o\prime})}{RT}\right]}{RT\left[1 + (b_{ox}/b_{rd}) \exp\left[-\frac{nF(E-E^{o\prime})}{RT}\right]\right]^2} \tag{3.8}$$

In this equation, the b_{ox}/b_{rd} ratio represents the relative adsorption strengths of the oxidized and reduced species. For an irreversible process [79],

$$i = \mathrm{nFAk\Gamma exp}\left[-\frac{\alpha \mathrm{n_a F(E-E^{o\prime})}}{\mathrm{RT}}\right]\exp\left[\frac{\mathrm{kRT}}{\alpha \mathrm{n_a Fv}}\exp\left[-\frac{\alpha \mathrm{n_a F(E-E^{o\prime})}}{\mathrm{RT}}\right]\right] \tag{3.9}$$

Equations 3.8 and 3.9 tend to a linear dependence of $\ln i$ on E for the limiting case when $\mathrm{E} \gg \mathrm{E^{o\prime}}$; i.e., at the beginning of the voltammetric peak.

Extension of this treatment to pulse techniques can, in principle, be made for several cases. In the case of square-wave voltammetry, theoretical current-potential curves for reversible electron transfer between species in solution are given by [184, 185]

$$\Delta i = \frac{nFAc\sqrt{Df}}{\sqrt{\pi}} \sum_{m=1}^{j} \frac{(-1)^j (Q_m - 2Q_{m-1} + Q_{m-2})}{(\delta + j - m)^{1/2}} \tag{3.10}$$

Where

$$Q_j = \frac{\exp(E_j - E^{o\prime})(nF/RT)}{1 + \exp(E_j - E^{o\prime})(nF/RT)} \tag{3.11}$$

In these equations, E_j represents the electrode potential during the jth half period, δ the fraction of the square-wave half period at which the current is measured, f is the square-wave frequency (equal to the inverse of the square-wave period), and the other symbols have their customary meaning. As long as the square-wave amplitude, E_{SW}, is lower than $0.5RT/nF$—a condition easily accomplished under the usual experimental conditions—the differential sum of the currents flowing during the anodic and cathodic half cycles can be represented by an expression such as [184]

$$\Delta i = H \frac{\exp(E - E^{o\prime})(nF/RT)}{[1 + \exp(E - E^{o\prime})(nF/RT)]^2} \tag{3.12}$$

Here, H represents an electrochemical constant.

Equation 3.12 again reduces to a Tafel linear $\ln i$ on E dependence in the foot of the voltammetric peak, a situation that applies with reasonable approximation for square-wave voltammograms of surface-confined species [79, 131].

Assuming that solid-state electrochemical processes involved in our voltammetry of microparticles analysis satisfy Tafel dependence between current and potential at the rising portion of voltammetric curves, the current can be approached by the expression

$$\mathrm{i} \approx \mathrm{q_o k}\exp\left(-\frac{\alpha \mathrm{n_a F}}{\mathrm{RT}}\mathrm{E}\right) \tag{3.13}$$

where q_o represents the total charge involved in the complete reaction of the electroactive solid. In the following section, it will be assumed that this kind of relationship is applicable not only to linear potential scan techniques, but also to pulse techniques, but also to pulse ones, in particular to square wave voltammetry, merely by inserting a numerical 'electrochemical' constant.

Equation 3.13 predicts a linear dependence of $\ln i$ on E whose slope depends on the coefficient αn_a, while the ordinate at the origin depends on the electrochemical rate constant and the net amount of depolarizer deposited on the electrode. Accordingly, both the slope and the ordinate at the origin of Tafel plots become phase-dependent [133, 183]. Since the quantity of depolarizer varies from one experiment to another, it is convenient to use normalized currents. For this purpose, one introduces the expression for the peak current for a linear scan voltammetric current:

$$i_p = B\left(\frac{\alpha n_a F}{RT}\right) v q_o \tag{3.14}$$

With B being an electrochemical coefficient of the response characteristic of the electrochemical process, the electrode area, and v being the potential scan rate, combining Eqs. 3.13 and 3.14 obtains

$$\ln(i/i_p) = \ln\left(\frac{kRT}{H\alpha n_a F}\right) - \frac{\alpha n_a F}{RT}E \tag{3.15}$$

Here, both the Tafel slope ($SL = \alpha n_a F/RT$) and the ordinate at the origin ($OO = \ln(kRT/B\alpha n_a F)$) become characteristic of the solid, but are independent on the amount of analyte deposited on the electrode.

For a material containing a mixture of microparticles of two electroactive analytes that display highly overlapped voltammetric peaks, the currents at the beginning of the common voltammetric wave can be taken as additive, assuming that there is no significant interaction between the compounds. Accordingly, one can write

$$i \approx q_{oX} k_X \exp\left(-\frac{\alpha_X n_{aX} FE}{RT}\right) + q_{oY} k_Y \exp\left(-\frac{\alpha_Y n_{aY} FE}{RT}\right) \tag{3.16}$$

If $\alpha_j n_{aj} FE/RT << 1 (j = X, Y)$, one can use the approximation $e^{-z} \approx 1 - z$, so that Eq. 3.15 reduces to:

$$i \approx (q_{oX} k_X + q_{oY} k_Y) \exp\left[-\frac{(q_{oX} k_X \alpha_X n_{aX} + q_{oY} k_Y \alpha_Y n_{aY})(FE/RT)}{q_{oX} k_X + q_{oY} k_Y}\right] \tag{3.17}$$

Assuming as before that voltammetric peaks for X and Y are strongly overlapped, a unique peak will be recorded, and the peak potential can be expressed as:

$$i_p \approx B_X \left(\frac{\alpha_X n_{aX} F}{RT}\right) v q_{oX} + B_Y \left(\frac{\alpha_Y n_{aY} F}{RT}\right) v q_{oY} \tag{3.18}$$

Thus the i/i_p ratio will be given by the approximate expression:

$$\frac{i}{i_p} \approx \frac{(q_{oX} k_X + q_{oY} k_Y) RT}{(B_X \alpha_X n_{aX} q_{oX} + B_Y \alpha_Y n_{aY} q_{oY}) nFv} \exp\left[-\frac{(q_{oX} k_X \alpha_X n_{aX} + q_{oY} k_Y \alpha_Y n_{aY})(FE/RT)}{q_{oX} k_X + q_{oY} k_Y}\right] \tag{3.19}$$

This equation again fits to a linear dependence of $\ln i$ on E. This means that the sample containing a mixture of X plus Y should give a linear Tafel plot. Remarkably, both the slope and the ordinate at the origin of that representation should be intermediate between those obtained for the X and Y components separately (Eq. 3.15).

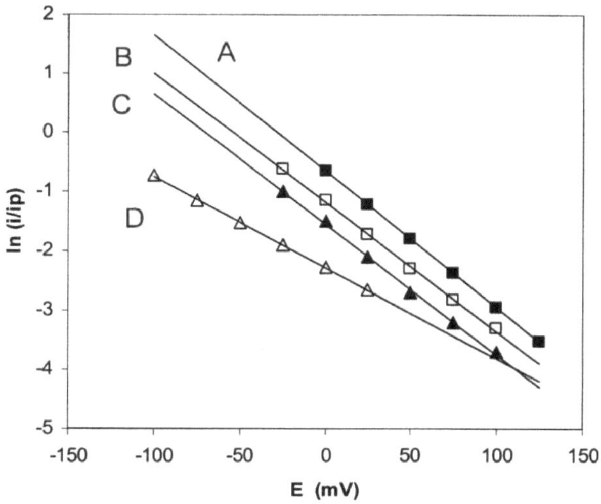

Fig. 3.11 Tafel plots for verdigris (**A**), atacamite (**B**), paratacamite (**C**), and cuprite (**D**) from linear scan voltammograms at sample-modified, paraffin-impregnated graphite electrodes immersed in 0.50 M potassium phosphate buffer (pH 7.0). Potential scan rate 50 mV/s

Typical Tafel plots for different copper materials are shown in Fig. 3.11. In all cases, an excellent linearity was obtained for $\ln(i/i_p)$ on E representations in terms of the correlation coefficient for linear fitting. Similar results were obtained for binary or ternary mixtures of such materials where highly overlapping peaks were recorded, both using linear potential scan and square-wave voltammetries.

This similarity makes it difficult to use pattern recognition criteria described in the precedent section because the involved parameters are not additive for mixtures of compounds, so that Tafel analysis of the rising portion of voltammetric curves offers a plausible alternative for resolution of mixtures. This is prompted by the fact that the values of *SL* and *OO* fell in well-separated regions in two dimensional diagrams, as shown in Fig. 3.12.

An example of the application of Tafel analysis is provided by samples taken for a bronze *montefortino* helmet from the Cabriel river valley (Kelin and Ikalesken period) in the Valencian region of Requena, dated back to the Second Iron Age (see Fig. 3.13). Upon attachment to paraffin-impregnated graphite electrodes immersed in 0.50 M phosphate buffer, voltammetric signals such as depicted in Fig. 3.14 were found [183].

According to the previous treatment, linear Tafel plots of $\ln(i/i_p)$ on E were obtained for samples providing from different regions of the helmet. As can be seen in Fig. 3.15, 2D diagrams using Tafel slope (*SL*) and Tafel ordinates at the origin (*OO*) yield a distribution of samples in three groups: one formed by cuprite (CI-11), one formed by atacamite (CI-4, CI-6, CI-9), and one formed by a mixture of cuprite plus atacamite (C-12).

A similar situation was found in the previously described frescoes of the vault of the *Sant Joan del Mercat* church in Valencia (Spain). In several regions of the vault, the attribution to this painter was uncertain because it was documented

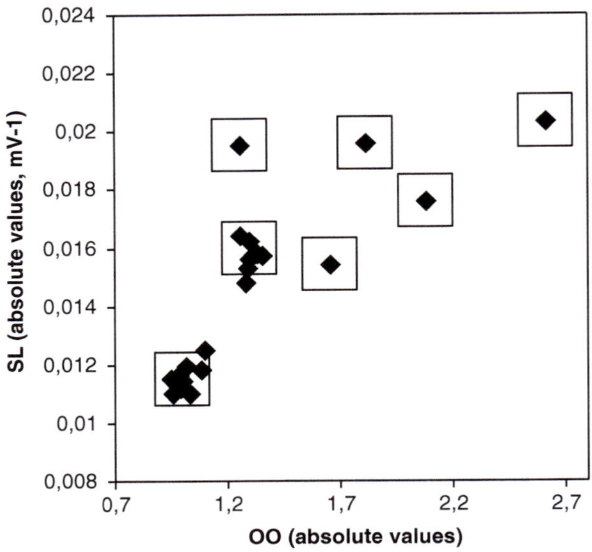

Fig. 3.12 Tafel *SL* vs. *OO* diagram for the most common copper pigments and copper alteration products. From SQWVs of specimen-modified paraffin-impregnated graphite electrodes immersed into 0.50 M phosphate buffer, pH 7.4. Potential scan initiated at +0.45 mV in the negative direction. Potential step increment 4 mV; square wave amplitude 15 mV; frequency 2 Hz

Fig. 3.13 Montefortino helmet from the Cabriel river valley (Kelin and Ikalesken period) in the Valencian region of Requena, dating back to the Second Iron Age [183]

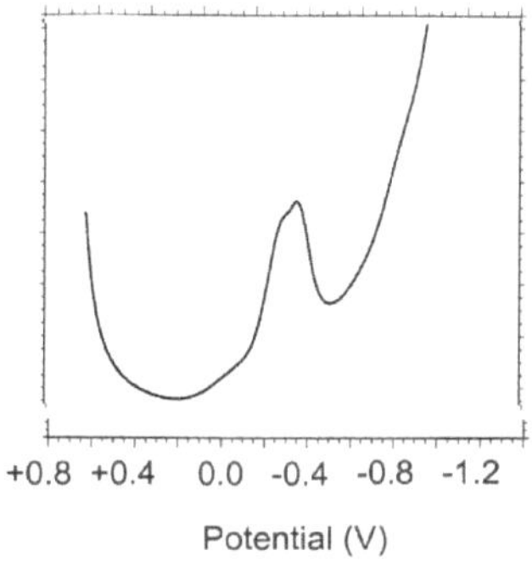

Fig. 3.14 Square-wave voltammograms of PIGEs modified with sample CI-12 from a bronze *montefortino* helmet from the Cabriel river valley (Kelin and Ikalesken period) in the Valencian region of Requena, dating back to the Second Iron Age. Electrolyte: 0.50 M potassium phosphate buffer, pH 7.0. Potential scan initiated at +650 mV in the negative direction. Potential step increment 4 mV; square wave amplitude 25 mV; frequency 5 Hz

that, in such regions of the vault, painter Vicente Guilló Barceló (1645–1698) started to execute a prior frescoes that were partially maintained despite the fact that Antonio Palomino was the painter in charge of the painting of the complete vault.

SEM/EDX analysis indicated that copper, frequently accompanied by cobalt pigments, was present in the samples from such frescoes. Analysis of voltammetric curves indicated that azurite or malachite, eventually accompanied by smalt, were the original pigments, and were accompanied (as previously described) by tenorite

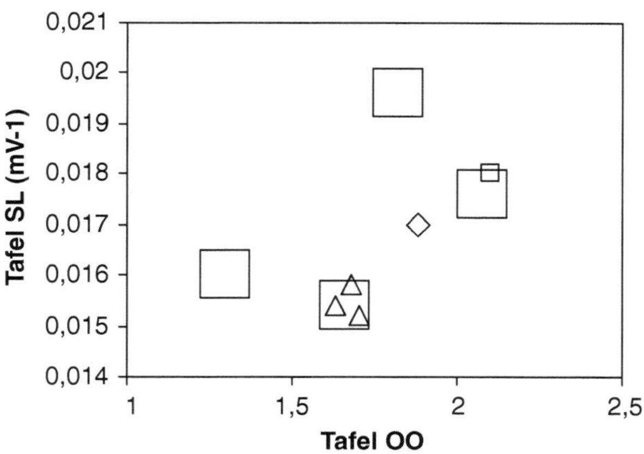

Fig. 3.15 Two-dimensional (2D) Tafel diagram corresponding samples CI-11 (*square*), CI-4, CI-6, CI-9 (*triangles*), and CI-12 (*rhomb*) from a bronze *montefortino* helmet from the Cabriel river valley (Kelin and Ikalesken period) in the Valencian region of Requena, dating back to the Second Iron Age. From SQWV data of sample-modified graphite electrodes in contact with 0.50 M phosphate buffer. Potential scan initiated at +0.65 V in the negative direction. Potential step increment 4 mV; square wave amplitude 25 mV; frequency 5 Hz. Adapted from [183]

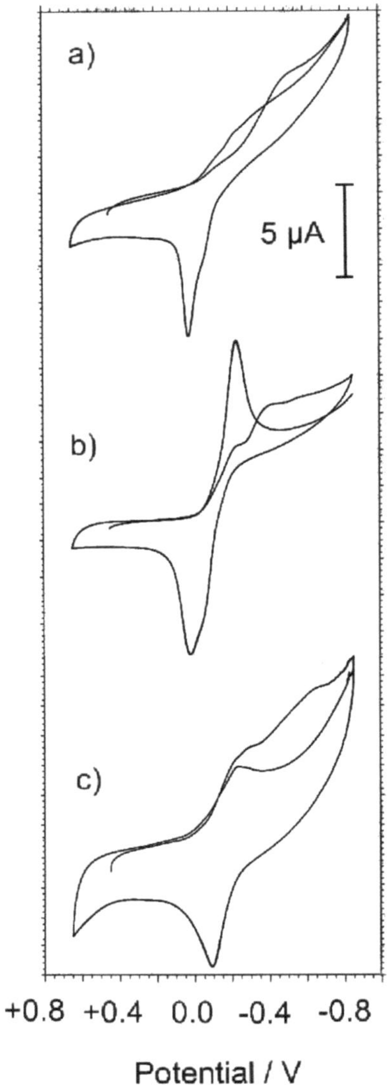

Fig. 3.16 CVs of PIGEs modified with (**a**) azurite, (**b**) malachite, and (**c**) smalt in contact (all Kremer pigments) immersed into 0.50 M potassium phosphate buffer of pH 7.4. Potential scan rate 50 mV/s [133]

in the zones exhibiting gunfire damage. Figure 3.16 shows the cyclic voltammetric responses of: (a) azurite, (b) malachite, and (c) smalt in contact with a phosphate buffer. In all cases, the reduction occurs at potentials ca. −0.10 V, leading to the corresponding metal deposit subsequently reduced via typical stripping processes at potentials ca. +0.05 V. As previously noted, analysis of stripping curves enables a clear distinction between copper and cobalt, thus indicating possible copper plus cobalt mixtures, as commented in Chap. 2 (see Fig. 2.10).

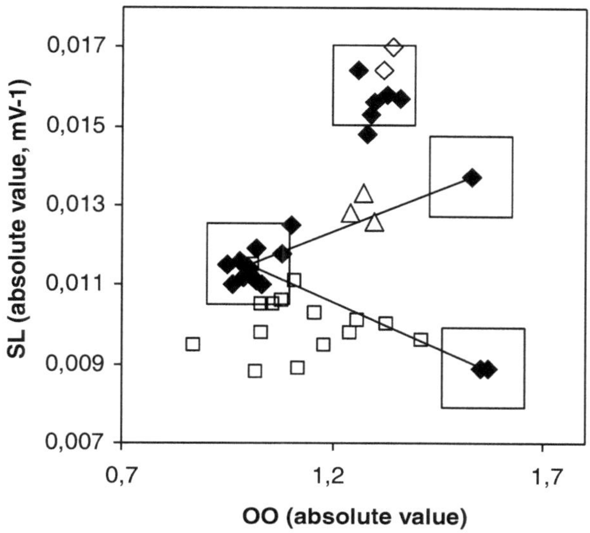

Fig. 3.17 A 2D Tafel slope vs. Tafel ordinate at the origin diagram for samples from the *Sant Joan del Mercat* church. From SQWVs at specimen-modified PIGEs immersed in 0.50 M phosphate buffer, pH 7.4 initiated at +0.65 V in the negative direction. Potential step increment 4 mV; square-wave amplitude 20 mV; frequency 5 Hz. *Solid figures* correspond to blanks of azurite, malachite, smalt, and smalt heated at 600°C. *Empty figures* correspond to samples provided from lightly altered regions of the vault (*squares*), strongly altered regions (*triangles*), and regions of the frescoes whose attribution to Palomino was uncertain (*rhombs*) [133]

In view of the close vicinity of the responses of these pigments, Tafel analysis was used for analyzing samples from the *Sant Joan del Mercat* church. The results are shown in Fig. 3.17, where samples are grouped into three types: those extracted from lightly and highly altered (blackened) areas of the vault in regions whose attribution to Palomino was clear, and those extracted from areas of the vault whose attribution to Palomino was in question. Blank specimens from commercial azurite, malachite, and smalt pigments, and those submitted to heating in a furnace at temperatures between 200 and 600°C were used. Interestingly, the heating of azurite and malachite specimens produced tenorite with no significant alteration in the Tafel parameters of surviving azurite or malachite signals. In contrast, heated specimens of smalt produced a significant shift in Tafel parameters [133].

As can be seen in Fig. 3.17, samples attributed to Palomino are divided into two regions between azurite and unheated smalt, and azurite and heated smalt—clearly denoting that Palomino used mixtures of both pigments in his paintings. Remarkably, the two samples from an uncertain author areas fell within the region of malachite. These results clearly suggest that these samples should be attributed to Vicente Guilló rather than to Palomino [133].

3.5 Multivariate Methods

Multivariate chemometric methods have claimed considerable attention in the last few decades because of their inherent capacity for resolving multicomponent, complex systems. Applications of multivariate methods in different electrochemical techniques have been recently reported by several authors [192–194].

An example of the application of these methods in the electrochemical field is the identification of earth pigments in paint layers. "Earths" are iron-based pigments composed by iron oxides and silicates. On the basis of their hue, iron earths can be divided into ochres (yellow), earths (red), siennas, umbers and green earths. On the basis of their mineralogy, earth pigments can be divided into clay-like yellow earths, calcium sulphate-based earths, umbers, and iron silicates. The most abundant clay-like earths results from lateritic weathering of mafic rocks, that are mainly composed by Al-substituted goethite (and/or hematite), kaolinite, and quartz. Gesso-type yellow and red earths are composed of goethite and/or hematite accompanied by gypsum, anhydrite, and clay minerals. These earths were probably of evaporitic origin. Dark earths are goethite-based pigments, probably originating from the weathering of iron minerals such as iron sulfides, whereas green earths contain iron silicates such as glauconite and celadonite. Finally, several pigments such as Venetian red are manufactured materials [137].

Earths are therefore multicomponent materials that contain not only iron oxides, but also a variety of minerals—namely, kaolinite, quartz, gypsum, anhydrite, and calcite, among others. The hue of ochres and raws is due to the absorption associated with the charge transfer between the ligand (OH^- and O^{2-}) and the Fe^{3+} ion contained in goethite and/or hematite, and depends significantly on the shape and size distribution of particles [195]. As a result of this complex composition, data analysis in techniques such as XRD and FTIR is made difficult by the coexistence of numerous and strongly overlapping signals due to the different components in the sample.

Figure 3.18 illustrates typical square-wave voltammograms of earth pigments attached to graphite electrodes in contact with 0.10 M HCl. Voltammetric curves can, in principle, be grouped into five morphological types whose "holotypes" are depicted in Fig. 3.18 [139, 174]. These holotypes correspond to: (a) Spanish hematite, (b) French yellow ochre, (c) sienna raw, (d) Italian toast umber, and (e) greenish natural umber [139, 174].

The voltammetric response of hematite-based pigments consists of a main cathodic peak at −0.40 V. For French ochres, peaks at −0.15 and −0.40 V are recorded, while siennas' square-wave voltammograms show overlapping peaks at −0.15 and −0.25 V followed by a shoulder near −0.40 V. For umbers, a well-defined peak at +0.25 V precedes a voltammetric profile similar to that of siennas. Finally, for greenish natural umber, the voltammogram is essentially limited to a prominent peak near +0.25 V, preceded by a peak at +0.58 V. This last can be attributed to the presence of MnO_2—a frequent component of earths, whose electrochemistry has been widely studied and described [113, 128, 171, 196–202]. The observed peak can be described in terms of the reduction process:

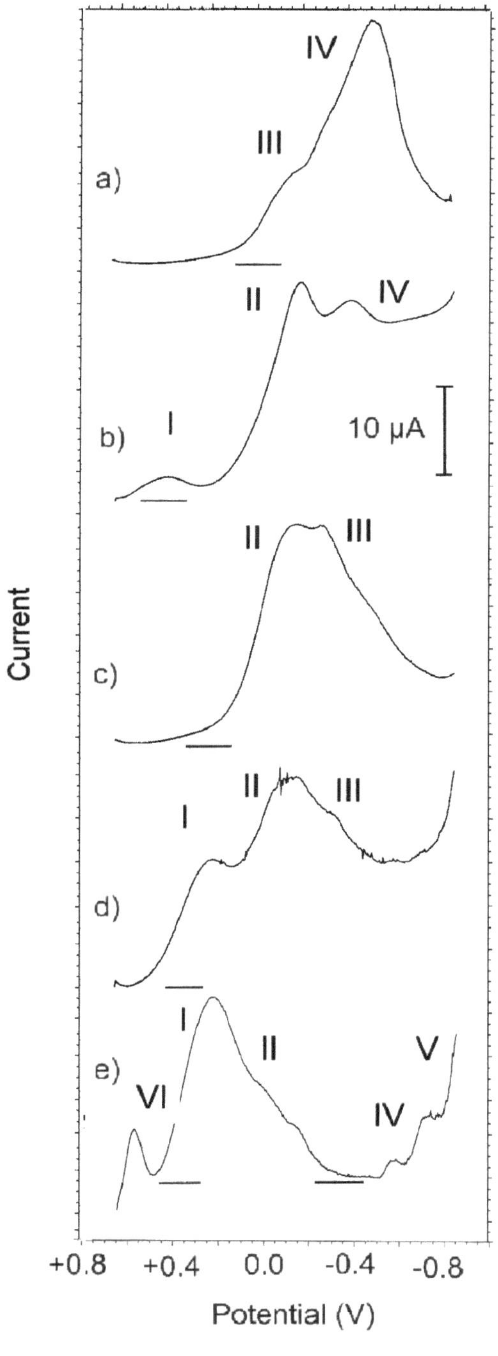

Fig. 3.18 Square-wave voltammograms of PIGEs modified with (**a**) Spanish hematite, (**b**) *yellow French ochre*, (**c**) sienna raw, (**d**) Italian toast umber, and (**e**) *greenish* natural umber immersed in 0.10 M HCl. Potential scan initiated at +0.65 V in the negative direction; potential step increment 4 mV; square wave amplitude 25 mV; frequency 5 Hz [139]

$$MnO_{2s} + H^{+}{}_{aq} + e^{-} \cdots \rightarrow MnO(OH)_s \qquad (3.20)$$

The electrochemistry of iron(III) oxides, hydroxy-oxides, and related materials has received attention in the last few decades [108, 113, 114, 137–147]. The electrochemical parameters are phase-specific and the kinetics of the electrochemical reaction are affected by the pH and the presence of other chemisorbing ions and/or complexing species. The kinetics of the electrochemical process can be described on the basis of the model developed by Grygar [147, 148]. Accordingly, the rate of reductive dissolution is driven by the detachment, via ion diffusion or complexation reaction, of metal centers from the reduced metal sites in the surface of the solid particles. As a result, the position and shape (roughly, peak width) of the voltammetric peaks depend on the average particle size and the homogeneity of the particle size distribution [147, 148]. Thus, voltammetric peaks are shifted cathodically by increasing Al-for-Fe substitution, increasing particle size, and anodically by increasing departures of stoichiometry—the wider peaks hence indicate less consolidated, and hence probably more heterogenous, iron oxides [137].

Electrochemical identification of earths can be derived from shape-characterizing parameters. Roughly, peaks at +0.25 V (I) can be attributed to amorphous, highly hydrated iron oxide forms, whereas peaks at −0.15 V (II) and −0.30 V (III) are attributable to goethite and hematite forms with variable degrees of hydration and crystallinity. Peaks at −0.40 V (IV) can unambiguously be attributed to crystalline hematite forms. For chemometric purposes, normalized peak currents, $I(j)$, for the above peaks (j = I, II, III, IV) were tabulated and included in 2D diagrams. As can be seen in Fig. 3.19, where the $I(II)/I(IV)$ ratio is plotted against the $I(III)/I(IV)$

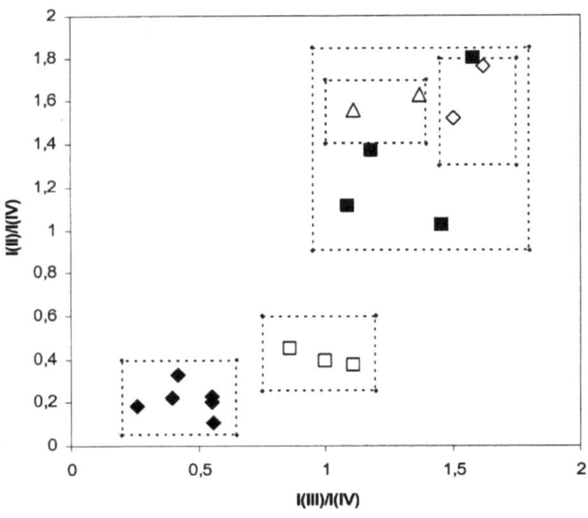

Fig. 3.19 Plot of $I(II)/I(IV)$ vs. for earth pigments differentiating hematite-based (*solid rhombs*), goethite-based clays (*solid squares*), siennas and Spanish ochre (*rhombs*), French ochres (*squares*), and umbers (*triangles*). From square-wave voltammograms of pigment-modified PIGEs immersed into 0.10 M *HCl*. Potential scan initiated at +0.65 V in the negative direction; potential step increment 4 mV; square wave amplitude 25 mV; frequency 5 Hz [139]

Table 3.2 Peak current ratios for commercial iron oxide (Fe_2O_3, Aldrich) and earth pigments from Kremer. From square-wave voltammograms of sample-modified PIGEs immersed into 0.10 M HCl. Potential step increment 4 mV; square wave amplitude 25 mV; frequency 5 Hz. Adapted from ref. [139]

Pigment	I(I)/I(IV)	I(II)/I(IV)	I(III)/I(IV)	I(VI)/I(IV)
Fe_2O_3FE	0.02	0.20	0.55	0.00
Hematite HE	0.03	0.33	0.42	0.00
Spanish hematite, SE	0.03	0.23	0.55	0.00
Caput mortum, CM	0.03	0.18	0.26	0.00
Venetian red, VR	0.05	0.22	0.40	0.00
Red Bole, RB	0.24	1.37	1.18	0.00
Red Clay, RC	0.76	1.90	1.60	0.00
Goethite GO	0.07	1.11	1.09	0.00
Spanish ochre, SO	0.17	1.02	1.46	0.00
Sienna raw, SR	0.31	1.81	1.62	0.00
Natural Sienna, NS	0.05	1.52	1.50	0.00
Yellow French ochre, YF	0.11	0.37	1.11	0.00
Orange French ochre, OO	0.23	0.39	1.00	0.00
French ochre, FO	0.21	0.45	0.86	0.00
Cyprus umber, CU	0.42	1.56	1.11	0.12
Italian toast umber, IU	1.00	1.63	1.37	0.14
Greenish natural umber	15.4	8.5	4.2	0.33

ratio, the diagram permits a clear distinction between earths, Spanish ochres, and French ochres [139, 174]. The diagram point for green earth was considerably separated from the above and has not been represented.

To obtain a detailed separation of the different types of pigments, and establish hidden relationships between the different voltammetric types, hierarchical cluster analysis was performed. Applying a Dendrogram building analysis of all data in Table 3.2 is shown in Fig. 3.20a. Here, greenish natural umber (GU) separates clearly from all other pigments (ERT branch) so that a second analysis could be performed for all remaining earths. The result is shown in Fig. 3.20b. Here, earth pigments are divided into two major branches. The first one is subdivided into hematite-based materials (FE, HE, SE hematites, Caput mortum, and Venetian red) and French ochres (FO, YF, OO). The second major branch of earths contains (roughly) goethite-based raw materials (GO, RB, RC), Spanish ochre (SO), siennas (SR, NS), and umbers (IU, CU).

3.6 Speciation

A speciation of particular interest is to detect the presence of species in different oxidation states in the sample. The determination of the presence of two or even more oxidation states of a given species, and eventually their relative quantitation, is a highly important analytical target in a variety of applications. The determination of the $Fe(III)/Fe(II)$ ratio in raw and ceramic materials is probably one of the most

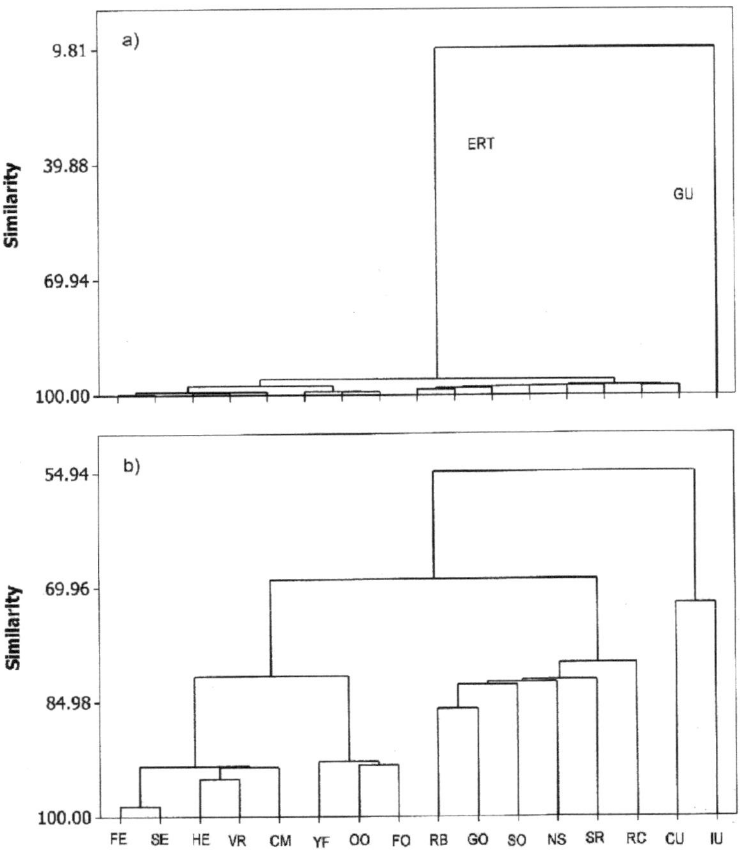

Fig. 3.20 Hierarchical cluster analysis with euclidean distance of the autoscaled variables applied to voltammetric parameters recorded for mineral and pigment specimens studied here. From data in Table 3.2: (**a**) including greenish natural umber; and (**b**) excluding this pigment [139]

paradigmatic examples of this kind of problem. This ratio provides information on the composition of raw materials but also on the nature of thermal treatments used for preparing ceramics.

The general problem of determining the relative amounts of oxidized and reduced forms of an electroactive species in solution was faced theoretically by Scholz and Hermes [203] for the cyclic voltammetry of an electrochemically reversible process controlled by diffusion. These authors used the currents at the larger and lower potential limits (anodic and cathodic switching potentials, $i_{\lambda,c}, i_{\lambda,a}$, respectively) represented in Fig. 3.21. Here, the CV is initiated at -500 mV in the positive direction of potentials, and it is assumed that the diffusion coefficients for the oxidized and reduced forms of the electroactive species (the depolarizer) are identical. Then, the

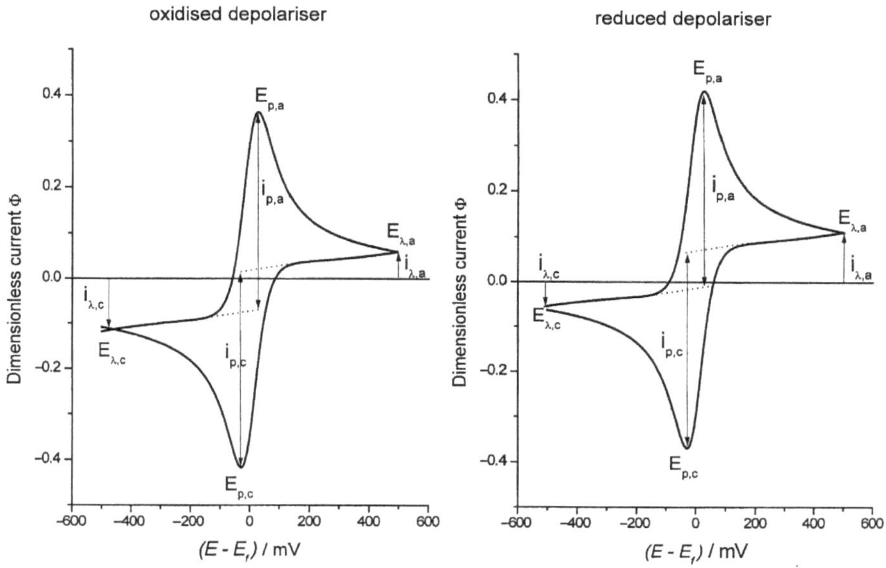

Fig. 3.21 Simulated cyclic voltammograms (2nd cycle) of an electrochemically reversible system with the depolarizer in the bulk of the solution that is in its reduced and oxidized forms, respectively taking: $E_{start} = E_{\lambda,c} = -500\,\text{mV}$, $D_{ox} = D_{rd} 10^{-5}\,\text{cm}^2/\text{s}$, scan rate 0.1 V/s [203]

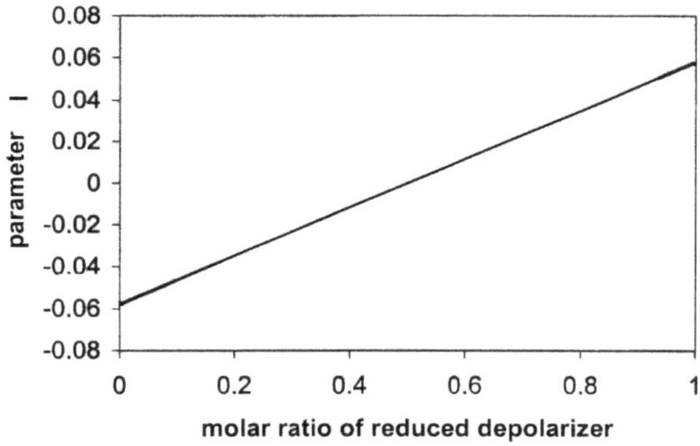

Fig. 3.22 Dependence of the *I* parameter on the molar ratio of the reduced depolarizer using simulated CVs in Fig. 3.21 [203]

parameter $I = |i_{\lambda,c}| - - |i_{\lambda,a}|$ varies linearly with the molar ratio of the reduced (or oxidized) form. That variation can be seen in Fig. 3.22.

A second procedure has also been proposed [204, 205]. This is based on the measurement of the peak currents for the anodic peak recorded in the first ($i_{pa}{}^*$) and the second (i_{pa}) scans when the voltammogram is initiated at the formal electrode potential of the couple in the positive direction of potentials, as illustrated

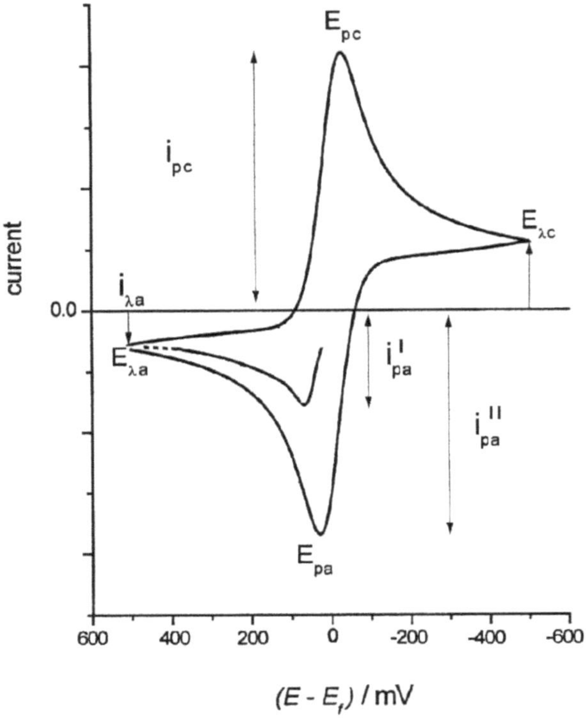

Fig. 3.23 Simulated cyclic voltammograms (1st and 2nd cycle) of an electrochemically reversible system with the depolarizer in the bulk of the solution in its reduced and oxidized forms, respectively taking: $E_{start} = E_{\lambda,c} = -500\,\text{mV}$, $D_{ox} = D_{rd}$ $10^{-5}\,\text{cm}^2/\text{s}$, scan rate 0.1 V/s. The potential scan is initiated at the formal electrode potential of that couple in the anodic direction. Adapted from [203]

in Fig. 3.23. Equivalently, cathodic peak currents can be used. The idea is that on initiating the potential scan for a reversible couple in the anodic direction, if the electroactive species is initially in its oxidized state, an anodic current initially flows so that $i_{pa}{}^*$ and i_{pa} will be similar. On the contrary, if the electroactive species is initially in its reduced form, the initial current will be cathodic, rapidly changing to anodic upon scanning the potential in the positive direction. As a result, the peak current recorded in the first scan will be significantly lower than that measured in the second and subsequent scans.

As a result, the values of the $(\text{i}_{\text{pa}}{}^*/\text{i}_{\text{pa}})$ ratio vary systematically with the potential scan rate and the molar fraction of reduced (or oxidized) form in the bulk of the solution. Pertinent data are shown in Fig. 3.24.

The use of these methods for speciation in solid materials requires that two essential assumptions—electrochemical reversibility and diffusive control—apply. Under these circumstances, theoretical CVs for ion-insertion solids are essentially identical to those for species in solution [206]. Since solid-state processes involve coupled

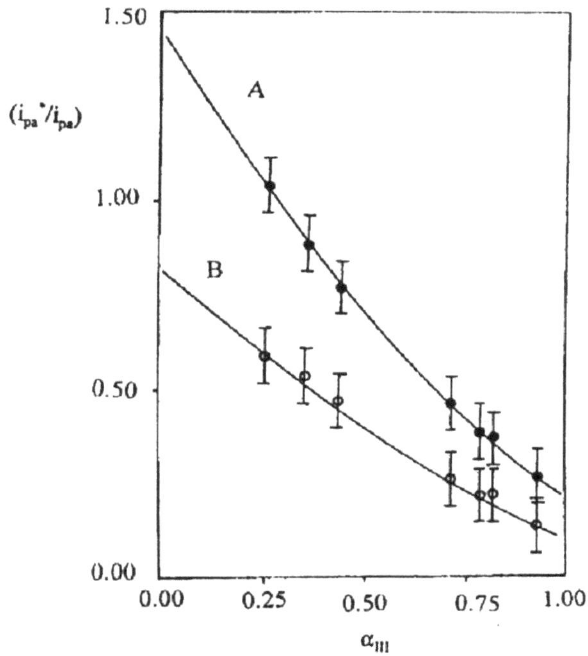

Fig. 3.24 Theoretical variation of the (i_{pa}^*/i_{pa}) ratio with the molar fraction of oxidized form in CVs for a one-electron reversible couple initiated at the formal electrode potential of that couple in the anodic direction for solutions containing different molar fractions of the oxidized form using the base line after (**A**) and before (**B**) the anodic peak [205]

diffusion of electrons and charge-balancing electrolyte ions through the solid, the condition of "ordinary" diffusive control requires relatively concentrated electrolyte solutions [206].

These methods have been applied to the determination of the Fe(III)/Fe(II) ratio in archaeological ceramic materials [205], and are of interest with regard to the so-called Maya Blue problem. Maya Blue is a famous artificial pigment widely used in murals, pottery, and sculptures by the ancient Mayas and other people in Mesoamerica. A typical Maya paint is shown in Fig. 3.25. The pigment hue ranges from bright turquoise to dark greenish blue [207]. A typical wall painting is illustrated in Fig. 3.26. The pigment exhibits a characteristic brightness and a remarkable chemical stability, and is unaffected by the attack of acids, alkalis, oxidants, reducing agents, organic solvents, or biodegradation.

Maya Blue can be described as a hybrid organic-inorganic material resulting from the association of indigo (H_2IN)—a blue dye extracted from *Indigofera suffruticosa* and other plants—with a local clay, palygorskite (known as attapulgite for historians and restorers), which is a fibrous phyllosilicate [207].

The nature of the indigo-palygorskite association, and the reasons for its hue and durability, have become controversial. Shepard [208] first introduced the idea of

Maya Blue as an unusual pigment in a dye attached to certain clays in Yucatan. In 1966, Van Olphen [209] prepared a complex analogous to Maya Blue by heating indigo with palygorskite or sepiolite—both clays with a fibrous structure. Pigments prepared with indigo plus laminar silicates were found not to be resistant to acidic attack. Van Olphen suggested that indigo molecules are too large to enter into the channels of the clay, so such channels are in some way sealed at their ends by indigo molecules [209]. Kleber et al., however, proposed that partial (or even deep) penetration of indigo molecules inside the palygorskite channels is possible [210]. More recently, José-Yacamán et al. [211, 212] proposed that iron and iron oxide nanoparticles in Maya Blue might at least partly account for the observed hue. In contrast, studies performed by Sánchez del Rio et al. [213, 214] did not find either iron in metallic form nor goethite in Maya Blue, in agreement with the idea that the blue coloration of the pigment is originated by the batochromic shift of the indigo absorption bands as a result of the association of the dye to the inorganic support [215–217].

Application of the voltammetry of microparticles techniques to Maya Blue samples, combined with FTIR and UV-VIS spectroscopies, electron microscopy (TEM, SEM/EDX), and atomic force microscopy (AFM) techniques, allowed Doménech et al. [218] to present a novel approach by characterizing the presence of dehydroindigo (the oxidized form of indigo) in Maya Blue [218] and determining thermochemical parameters for the attachment of indigo and dehydroindigo (and leucoindigo, the reduced form of indigo) to the palygorskite matrix [218, 219]. The electrochemical processes for indigo oxidation to dehydroindigo and indigo reduction to leucoindigo are shown in Scheme 3.1. These processes can be clearly recorded for indigo microparticles attached to graphite electrodes, in contact with aqueous electrolytes, as shown in Fig. 3.26a.

The effect of the indigo/dehydroindigo attachment to palygorskite on the Maya Blue hue is illustrated in Fig. 3.27, where the spectra of indigo, dehydroindigo,

Fig. 3.25 Maya blue painting from Calakmul, late classical period. Photograph by Mª Kuisa Vázquez de Agredos Pascual

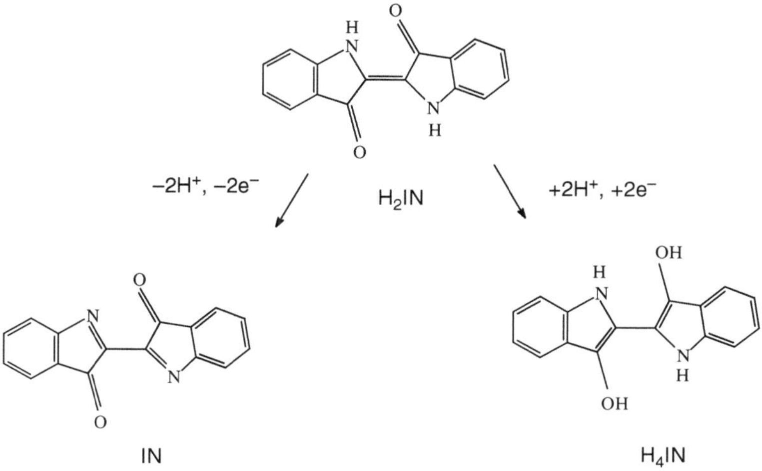

Scheme 3.1 Schematics for the electrochemical oxidation of indigo to dehydroindigo and the electrochemical reduction of indigo to leucoindigo [218]

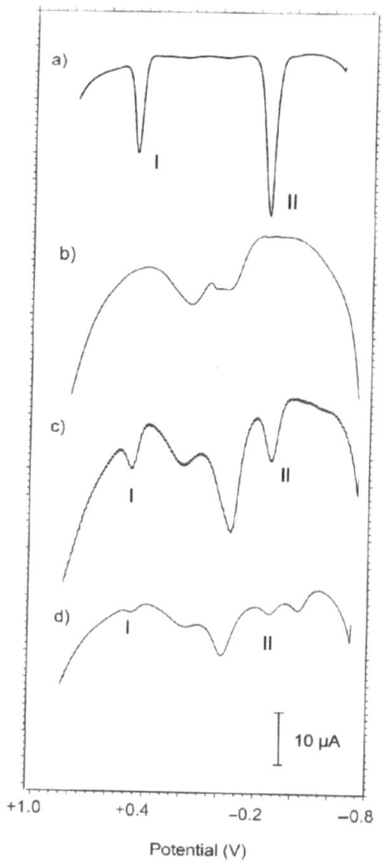

Fig. 3.26 SQWVs of PIGEs modified with (**a**) indigo and (**b**) pristine palygorskite from Sak lu'um, and MB samples from (**c**) Chakmultún, and (**d**) El Tabasqueño, immersed in 0.50 M HAc +0.50 M NaAc buffer at pH 4.85. Potential scan initiated at −750 mV in the positive direction. Potential step increment 4 mV; square-wave amplitude 25 mV; frequency 5 Hz [223]

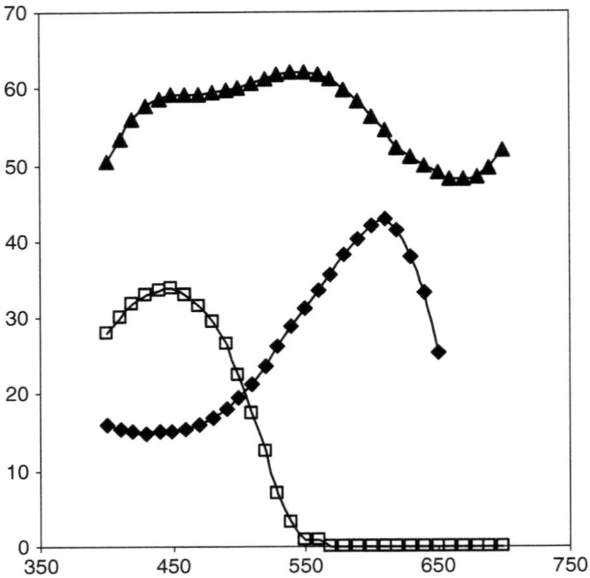

Fig. 3.27 Visible spectra for indigo (*rhombs*), dehydroindigo (*squares*), and a Maya Blue sample (*triangles*) from Calakmul, Late Preclassical period

and a typical Maya Blue samples are superimposed. As studied both theoretically and experimentally by Klessinger and Lüttke [220], indigo exhibits an absorption band with a maximum typically located at 605 nm, while dehydroindigo yields an absorption band with maximum at 425 nm resulting, respectively, in blue and yellow colorations. Spectra of Maya Blue samples produced a two-peak spectrum with maxima located at 440 and 570 nm, consistent with the blue-green coloration of the pigment.

Speciation techniques previously described allowed for estimating the mean dehydroindigo/indigo ratio in Maya Blue samples, as well as its variation with the depth in palygorskite crystals [221]. A maximum dehydroindigo proportion of 25–30% can be established from electrochemical measurements [221].

Application of chemometric techniques to Maya Blue samples from different archaeological sites in Yucatán and Campeche (Mexico) led to propose that different preparation procedures of the pigment were used by the ancient Mayas, the recipes evolving in a ramified form [222]. Reproduction of ancient procedures for preparing indigo specimens from leaves of *Indigofera suffruticosa* [223] also suggested an evolution over time of preparation methods. Additionally, electrochemical data denoted the presence of the pigment in samples from wall paintings in the substructure II-C from the archaeological site of Calakmul, dated in the Late Preclassical period [218]—by far the most ancient sample of Maya Blue currently known.

Chapter 4
Quantitative Methods

4.1 Quantitation

Quantitation of components in samples is a general goal of analytical purposes. In the fields of archaeometry, conservation, and restoration, quantitation of analytes is of considerable interest in a wide variety of applications. Thus, the provenance of raw materials, alloy composition, and more, can be crucial for authentication, geographical location, or analysis of the time evolution of works of art, techniques, etc.

In conventional electrochemistry in solution, quantitation of analytes can be obtained by using several techniques. Thus, exhaustive electrolysis provides an absolute quantitation of an electroactive component in the sample. Voltammetric measurements (linear potential scan, cyclic, pulse, and square-wave techniques) can be used for determination of analytes in solution via calibration because peak currents (and peak areas) are usually proportional to the analyte concentration under fixed electrochemical conditions.

The use of such methods for electrochemical quantitation in conservation and restoration, however, should involve the attack of the sample by an adequate reagent and the subsequent determination of selected analytes in solution. This methodology involves handling of micro- or submicrosamples, and further treatment of highly diluted solutions with concomitant opportunities for sample loss and/or sample contamination.

In spite of its evident limitations, solid-state electrochemical methods can also be used for quantifying electroactive components in sparingly soluble solids. As discussed in this chapter, relative quantitative methods and absolute quantitations are available from voltammetric data.

The application of such methods requires that, under given electrochemical conditions (technique and parameters of the exciting signal, electrolyte), the involved electroactive species (the analyte(s) and reference compounds) produce well-defined signals whose "intensity" is proportional to the net amount of that species deposited on the electrode surface. The generalized "intensity" labeled here can correspond to peak current or peak area in voltammograms, as well as to other parameters/techniques such as charges in chronocoulometric experiments. Along this chapter, for brevity, we refer exclusively to voltammetric measurements.

A. Doménech-Carbó et al., *Electrochemical Methods in Archaeometry, Conservation and Restoration*, Monographs in Electrochemistry,
DOI 10.1007/978-3-540-92868-3_4,

4.2 Phase Composition

Electrochemical methods can be applied to the determination of the composition of solid phases as well as mixtures of solids [224–228]. The first situation is illustrated in Fig. 4.1, where cathodic voltammograms of *CuS, CuSe*, and a solid phase of composition $CuSe_{0.4}S_{0.6}$ reported by Meyer et al. [227] are shown. This last can be described as a solid solution formally regarded as a copper sulfide, in which 40% of sulfide ions have been replaced by selenide ions. The new phase produces a voltammetric peak at a potential intermediate between those for *CuS* and *CuSe*.

The mixed phase is reduced at a potential that depends on the molar ratio of the two salts formally forming the mixed crystal, as can be seen in Fig. 4.2, where peak potentials of different mixed crystals of $CuSe_{1-x}S_x$ are plotted against the molar ratio $x_S/(x_{Se}+x_S)$ [224]. Since the peak potentials vary slightly with the total amount of charge (Q) consumed in the electrochemical reaction, standardized peak potentials extrapolated to values for $Q = 0$ were taken.

Thus, Lamache and Bauer [100] determined the stoichiometry of different copper sulfides, using charge determination from the measurement of peak areas in cyclic voltammograms of chalcolite (Cu_2S) included in carbon paste electrodes. Chronocoulometric data have been systematically applied in combination with voltammetric measurements for determining the composition of solid phases by Scholz et al. [74–78]. As previously described (Chap. 2), in these experiments a constant potential was applied to the working electrode so that current/time and charge/time curves were obtained. The idea was that when electroactive species A, B, ... yielding separated electrochemical processes exist in the sample, the application of appropriate potential steps produces the consumption of charges $Q_A, Q_B, \ldots$ proportional, respectively, to the number of mol of the corresponding electroactive species in the sample, $x_A, x_B, \ldots$. Incorporating the number of electrons involved in the respective electrochemical processes, $n_A, n_B, \ldots$, the molar ratio of a selected pair of electroactive components can be obtained as:

$$x_A/x_B = (Q_A/Q_B)(n_B/n_A) \tag{4.1}$$

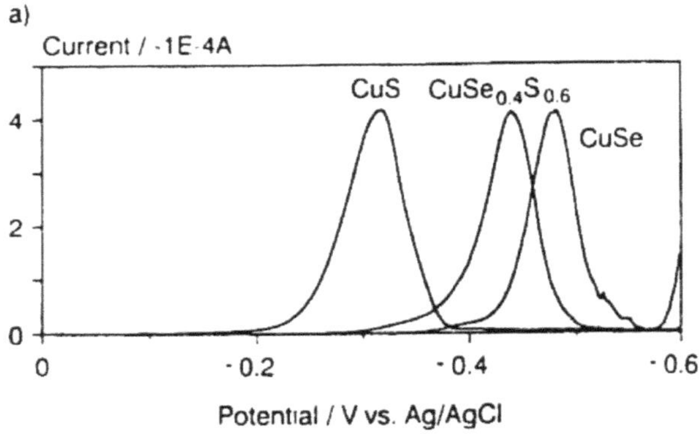

Fig. 4.1 Cathodic linear potential scan voltammograms of *CuS*, *CuSe*, and $CuSe_{0.4}S_{0.6}$ in 1 M H_2SO_4. Potential scan rate 11 mV/s [227]

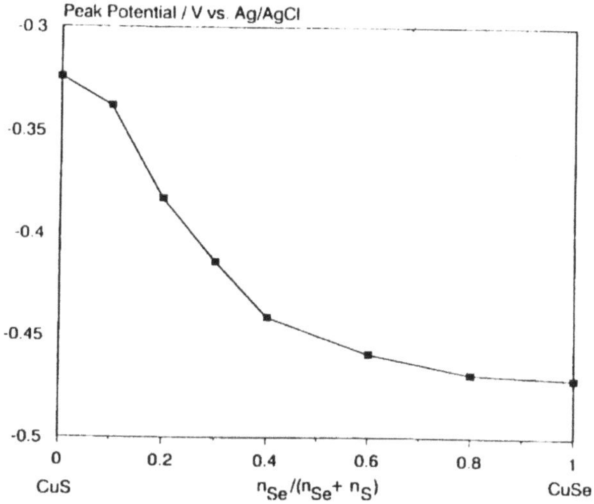

Fig. 4.2 Plot of standardized peak potentials (extrapolated to values for $Q = 0$) of different mixed crystals of $CuSe_{1-x}S_x$ vs. the molar ratio $x_S/(x_{Se}+x_S)$. From linear potential scan voltammograms in 1 M H_2SO_4 [227]

In most cases, the sample contains electrochemically depositable metals, so a scheme similar to that used in classical trace metal analysis in solution is applicable. Here, a first reductive step—consisting of the application of a sufficiently cathodic potential—is used. As a result, the metallic compound is reduced to the corresponding metal. After this electrodeposition (or electrogeneration) step, a detection step is used. In chronocoulometry, a sufficiently anodic potential for reoxidizing the metal deposit to the corresponding metal ions in solution is applied. In voltammetry, a potential scan in the positive direction is used, which is generally initiated at the electrodeposition potential used in the prior electrogeneration step. Here, sharp stripping peaks are obtained, thus providing high sensitivity.

Chronocoulometric data have been used for accurate determination of the composition of solid phases, as described by Scholz et al. [74–78, 224–228]. It should be noted, however, that coulometric measurements are always more demanding than voltammetric ones, because coulometry offers opportunity for more pronounced systematic errors. However, its use is highly recommended for cases where voltammetry fails [77, 78, 224].

4.3 Relative Quantitation

When two or more electrochemical species exist in the solid sample, their relative amounts can be estimated from electrochemical data using different procedures. Remarkably, when a mixture of solids is used, the voltammetric response looks like the superposition of the voltammograms of the parent solids. This can be seen in

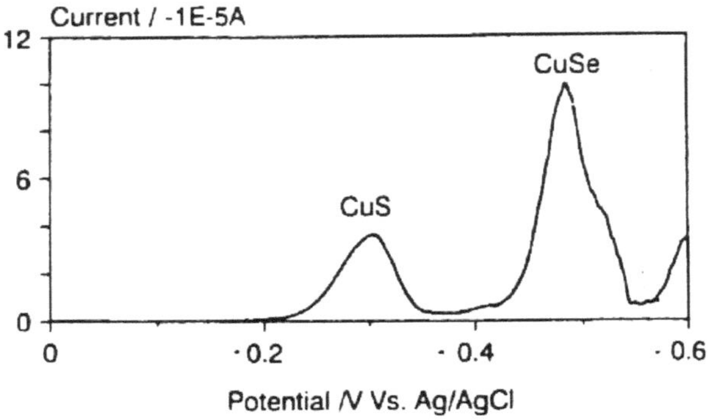

Fig. 4.3 Cathodic linear potential scan voltammograms of *CuS*, *CuSe*, and a mixture of *CuS* and *CuSe* (1:2 molar ratio) in 1 M H_2SO_4. Potential scan rate 11 mV/s [227]

Fig. 4.3, where voltammograms for *CuS*, *CuSe*, and a mixture of *CuS* and *CuSe* (1:2 molar ratio) are shown.

In the voltammetry of microparticles, relative quantitations can easily be obtained as follows. Let us first consider a sample containing two electroactive species, A and B. Assuming that under the selected electrochemical conditions (technique, parameters, electrolyte), separated signals are obtained for A and B, the voltammetric peak currents (or peak areas), i_A, i_B, can be taken as proportional to the number of mols of A and B, x_A, x_B, respectively transferred to the electrode. Then, one can write

$$i_A = G_A x_A = G_A m_A / M_A \quad (4.2)$$

$$i_B = G_B x_B = G_B m_B / M_B \quad (4.3)$$

where m_A, m_B represent the mass of A and B deposited on the electrode, and M_A, M_B represent their atomic or molecular mass. G_A, G_B represent the respective electrochemical coefficients of the response, depending on the technique, signal parameters (such as the potential scan rate in linear and cyclic voltammetry), and the characteristics of A and B and the electrolyte. It should be noted that, in general, G_A, G_B may be influenced by other electroactive or non-electroactive compounds in the sample. Assuming that this "matrix effect" is negligible, the relationship between the number of mols—or, equivalently, between masses of A and B in the electrode (which should be equal to that relationship in the original, well-homogenized sample)—can be estimated from the quotient between the peak currents, providing that the individual coefficients of response are known:

$$\mathrm{x_A/x_B = (G_A/G_B)(i_A/i_B)} \quad (4.4)$$

$$\mathrm{m_A/m_B = (G_A/G_B)(i_A/i_B)(M_B/M_A)} \quad (4.5)$$

This methodology has been applied for determining the relative composition of alloys [225], amalgams [226], and mixed crystals [227], among others [74–78], based on peak current measurements. The essential requisite is that both electroactive components behave independently—i.e., that the components of a mechanical mixture do not influence each other with respect to their thermodynamic activities in electrochemical reactions [77]. In Fig. 4.4, theoretical calibration plots for the absolute peak current, when the amount of mixture is constant for each measurement (left) and for the percentage peak current (right), are shown [77, 228].

In this context and more recently, Cepriá et al. quantified CdS in cadmium pigments (cadmiun sufide and cadmium sulfoselenide) by applying a reductive pretreatment followed by an oxidative scan of a sample plus AgCl (taken as a standard) mixture. The relative oxidation of the Cd oxidation peak relative to the oxidation of Ag was found to be linearly dependent on the Cd:Ag molar ratio in the mixture [229]. Grygar and van Oorschot used peak area measurements in linear-scan voltammograms for determining the limit of voltammetric detection of ferric oxides in soil samples [230]. Quantitation was done from peak-area measurements for the iron-centered peaks relative to that of an internal manganese oxide standard.

An illustration of relative quantitation procedures from peak current measurement applied to materials used in works of art is shown in Fig. 4.5, where square-wave voltammograms for Brazilwood and logwood, and different mixtures of both pigments attached to paraffin-impregnated graphite electrodes in contact with an aqueous phosphate buffer, are shown. Both dyes show a peak at ca. +0.20 V, which, for Brazilwood, is accompanied by a peak at +0.65 V. As can be seen in Fig. 4.6, variation of peak currents for such peaks with the percentage of Brazilwood fits to linear plots, in agreement with expectations (see Fig. 4.4). Notice that, in this case, absolute peak currents were used because identical amounts of the sample were mechanically transferred to the graphite electrode.

Peak areas can also be used for relative quantitation procedures. This is illustrated in Fig. 4.7, corresponding to the determination of the tenorite/(azurite + smalt) ratio

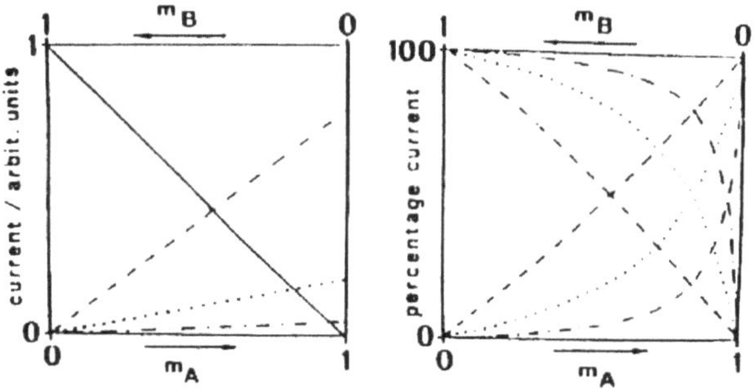

Fig. 4.4 Theoretical calibration plots for absolute peak current (*left*) in binary mixtures when the amount of mixture is constant for each measurement, and corresponding plots for the percentage peak current (*right*) [228]

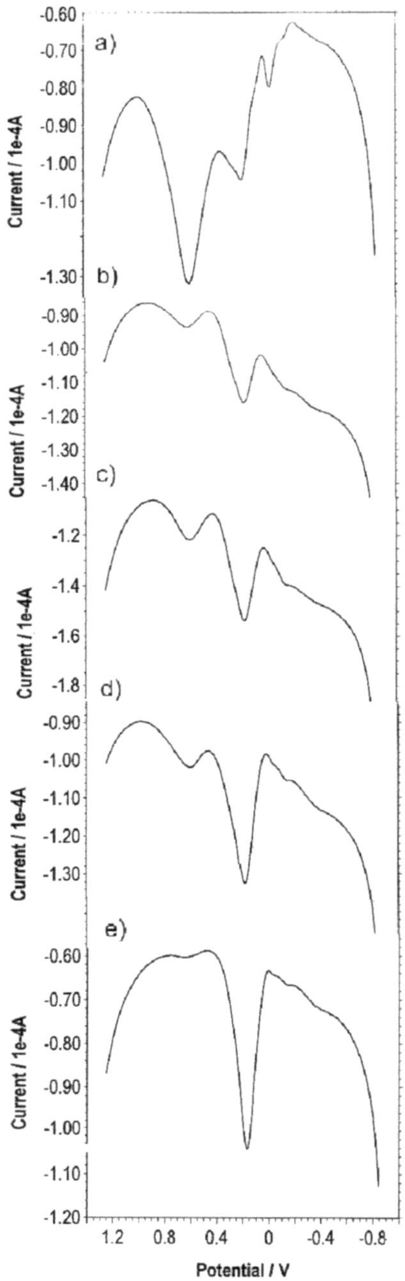

Fig. 4.5 Square-wave voltammograms for commercial Brazilwood (**a**) and logwood (**e**), and mixtures of both pigments containing 50% (**b**), 75% (**c**), and 85 % (w/w) (**d**) of Brazilwood, attached to paraffin-impregnated graphite electrodes. Electrolyte: 0.50 M phosphate buffer, pH 7.0. Potential step increment 4 mV; square-wave amplitude 25 mV; frequency 5 Hz

in thermally altered samples from the *Sant Joan del Mercat* church, using the quotient between the peak currents for voltammetric signals at −0.10 and −0.60 V (see Sect. 3.3) [133]. Since it was not possible to control the amount of sample transferred to the electrode surface, only relative values of peak areas or peak currents can be used.

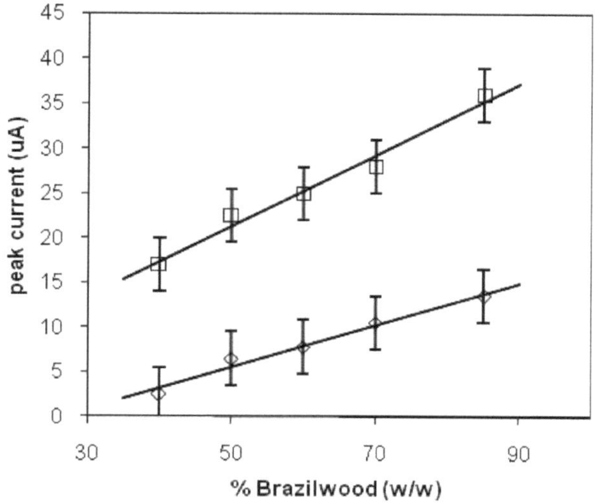

Fig. 4.6 Variation of the peak currents for peaks at +0.20 (*squares*) and at +0.65 V (*rhombs*) with the percentage of Brazilwood in square-wave voltammograms for Brazilwood plus logwood mixtures in contact with a phosphate buffer. Conditions are as in Fig. 4.5. Here, identical amounts of sample were transferred to the paraffin-impregnated graphite electrode

In the context of the analysis of cultural goods, this procedure has been used for the relative quantitation of metal ions in archaeological glass and glazes [231] using the aforementioned striping methodology. This can be seen in Fig. 4.8, where the square-wave voltammograms of cobalt blue Co_3O_4 (a,c), and a microsample from

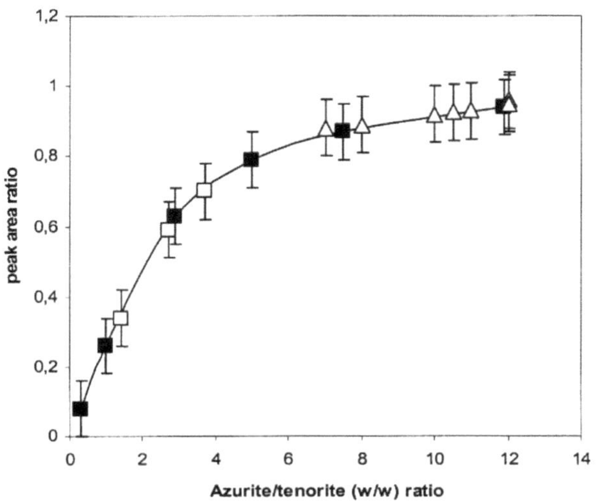

Fig. 4.7 Calibration graph for estimating the tenorite/(azurite + smalt) ratio in thermally altered samples from the *Sant Joan del Mercat* church, using the quotient between the peak currents for voltammetric signals at −0.10 and −0.60 V [133]

Fig. 4.8 Square-wave voltammograms of graphite-polyester composite electrodes modified by: cobalt blue (**a**,**c**), and glazed ceramic sample from Valencia, Spain (**b**,**d**), immersed into 0.10 M HCl. (**a**,**b**) Potential scan initiated at +0.65 V in the negative direction; (**c**,**d**) potential scan initiated at −1.05 V in the positive direction after an electrodeposition step of 60 s at −1.05 V. Potential step increment 4 mV, square-wave amplitude 25 mV, frequency 15 Hz [231]

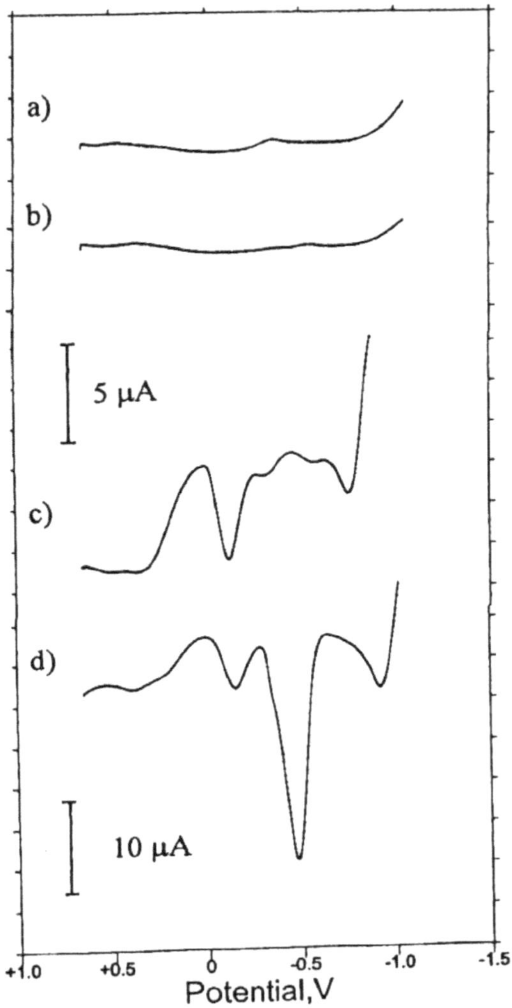

the glazed surface of a contemporary ceramic tile from Valencia (Spain) immersed into 0.10 M *HCl* (b,d) are shown. In voltammograms (a) and (b), the potential was scanned from +0.65 V in the negative direction, so that only weak reduction peaks were obtained for both the glazed sample and the standard material. Voltammograms (c) and (d) were initiated at −1.05 V, after the application of a constant potential of −1.05 V over 60 s. Here, well-defined stripping peaks were obtained at −0.45 V and −0.12 V, corresponding to the oxidation of the deposits of lead and cobalt metals, respectively, that were previously formed as a result of the reductive electrogeneration step.

Using voltammograms, as in Fig. 4.8, one can estimate the Co/Pb molar ratio in glazed ceramics from the quotient between the peak currents or the peak areas of the Co- and Pb-centered stripping processes. It should be emphasized, however, that the relative height of the involved stripping peaks (i.e., the G_A/G_B ratio) depends, in general, on the time and potential of the prior reductive electrodeposition step, as well as on the parameters of the stripping scan (initial potential and scan rate in linear-scan voltammetry, initial potential, potential step increment, square-wave amplitude, and frequency) in square-wave voltammetry, etc. Accordingly, prior calibration using materials with known relative amounts of cobalt and lead, and fixed electrochemical conditions, are demanded. Application of this methodology using electrochemical stripping voltammetry [232] and quartz microbalance measurements has been also reported [233].

Relative quantitation can also be obtained from voltammetric data in the case of strongly overlapped peaks using generalized Tafel analysis of the rising portion of voltammetric curves. For quantitation of a mixture of two components, X and Y, one can combine the Tafel dependence for individual components:

$$\ln(i/i_p)_Y = \ln\left(\frac{k_o RT}{H\alpha_Y n_{aY} F}\right) - \frac{\alpha_Y n_{aY} F}{RT} E \tag{4.6}$$

With that for the X plus Y mixture,

$$\ln(i/i_p) \approx \ln\left(\frac{(q_{oX}k_X + q_{oY}k_Y)RT}{(H_X\alpha_X n_{aX} q_{oX} + H_Y\alpha_Y n_{aY} q_{oY})nFv}\right) - \frac{(q_{oX}k_X\alpha_X n_{aX} + q_{oY}k_Y a_Y n_{aY})(FE/RT)}{q_{oX}k_X + q_{oY}k_Y} \tag{4.7}$$

As a result, the X to Y molar ratio, $f(= q_{oX}/q_{oY})$, can be expressed as

$$f = \left(\frac{SL_Y - SL_M}{SL_M - SL_X}\right)\left(\frac{k_Y}{k_X}\right)\left(\frac{\alpha_Y n_{aY}}{\alpha_X n_{aX}}\right) \tag{4.8}$$

In this equation, SL_M represents the Tafel slope for the mixture of X plus Y, and S_{LX}, S_{LY} represents the Tafel slopes for the individual components. This equation enables a determination of f from Tafel representations, providing that the quotients between the individual electrochemical rate constants, k_X and k_Y, and the electron transfer coefficients, $\alpha_X n_{aX}, \alpha_Y n_{aY}$, are known.

Considering Eq. 4.7, these ratios can be directly obtained from the normalized Tafel ordinates at the origin, OO_X, OO_Y, and the Tafel slopes, SL_X, SL_Y, for the individual components, so that finally,

$$f = \left(\frac{SL_Y - SL_M}{SL_M - SL_X}\right)\left(\frac{\exp(OO_Y)}{\exp(OO_X)}\right)\left(\frac{SL_Y}{SL_X}\right) \tag{4.9}$$

The application of this formalism requires, strictly, that the voltammetric peaks for both components coincide, and that X and Y behave electrochemically independent of each other. In most cases, however, although the voltammetric signals are strongly overlapped, there is no complete fit between them. As a result, the resulting peak broadens and the peak current in these mixtures will be lower than the sum of the peak currents for the separated components, thus distorting the i/i_p values with respect to those for exactly coincident voltammetric peaks. In these cases, Tafel diagrams can be modified depending on the separation between the peak potentials for the X and Y peaks. Thus, Fig. 4.9 compares theoretical working current-potential curves and experimental data for azurite plus smalt mixtures. The theoretical Tafel parameters were close to those experimentally determined for the corresponding mixtures that show a peak potential separation of 50 mV for voltammetric peaks for azurite and smalt [133].

This method was applied to samples from the Palomino's frescoes in the vault of the *Sant Joan del Mercat* church in Valencia, allowing to calculate the molar percentages of smalt relative to the azurite + smalt mixtures in such samples. Experimentally determined Tafel Parameters for such samples are compared in Fig. 4.10, with theoretical values of these parameters calculated for two voltammetric peaks having peak potential separations of 0, 50, 75, and 100 mV.

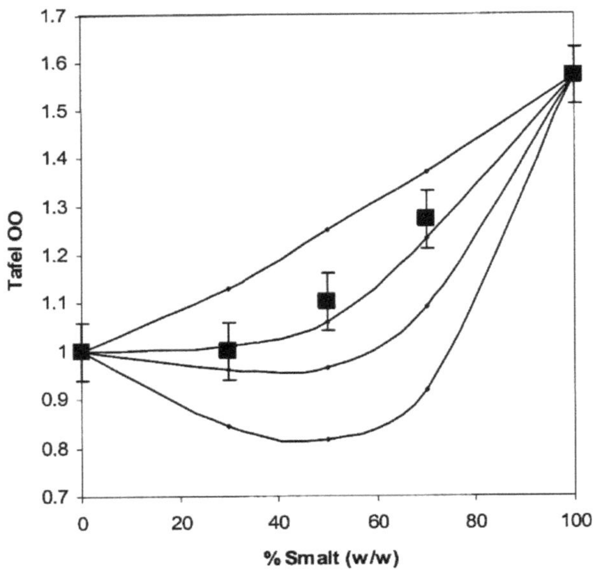

Fig. 4.9 Theoretical variation of the Tafel ordinate at the origin for azurite + smalt mixtures taking peak potential separations (from upper to below) of 0, 50, 75, and 100 mV. Data points correspond to synthetic specimens containing pure azurite, pure smalt, and 70:30, 50:50, and 30:70 (%, w/w) azurite-smalt mixtures. From SQWVs at specimen-modified PIGEs, immersed into 0.50 M phosphate buffer, pH 7.4 initiated at +0.65 V in the negative direction. Potential step increment 4 mV; square-wave amplitude 20 mV; frequency 5 Hz [133]

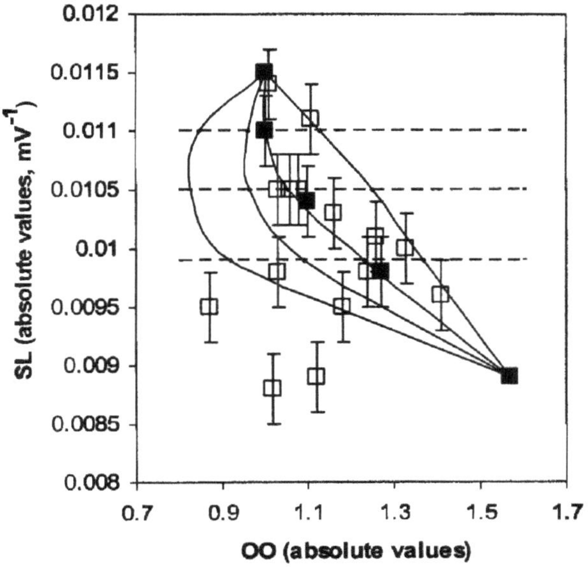

Fig. 4.10 Detail of the *SL* vs. *OO* diagram in the azurite + smalt region for samples from the *Sant Joan del Mercat church*. From SQWVs at specimen-modified PIGEs immersed in 0.50 M phosphate buffer, pH 7.4 initiated at + 0.65 V in the negative direction. Potential step increment 4 mV; square-wave amplitude 20 mV; frequency 5 Hz [133]

Tafel analysis of these mixtures indicated that pigmenting samples consisted of pure azurite, pure smalt, and azurite plus smalt mixtures, concentrated in smalt molar percentages of 55%, 72%, and 85%. This can be seen in Fig. 4.11, where a frequency diagram illustrative of the use of selected azurite/smalt dosages is presented.

Crossing all these data with the position of the samples in the nave provides a scene for the gunfire attack suffered by the paintings, illustrated by the variation of the percentage of tenorite with respect to the vault axis shown in Fig. 4.12. Accordingly, "blackened" samples with high tenorite content were placed along the central axis of the vault. Crossing the foregoing set of data with those derived from the analysis of earth pigments [139, 174], one can conclude that the central part of the vault reached temperatures between 600 and 650°C during the gunfire. Samples with minor amounts of tenorite provided from the lateral zones of the nave, probably experienced temperatures in the 350–460°C range. Samples from paintings near the lunettes, for which no significant tenorite signals were recorded, probably reached temperatures of about 260°C.

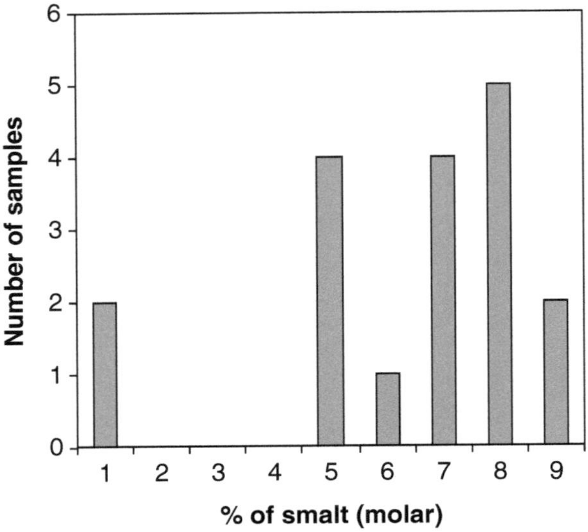

Fig. 4.11 Frequency diagram for the dosages of azurite + smalt mixtures determined in Palomino's frescoes in the *Sant Joan del Mercat* church

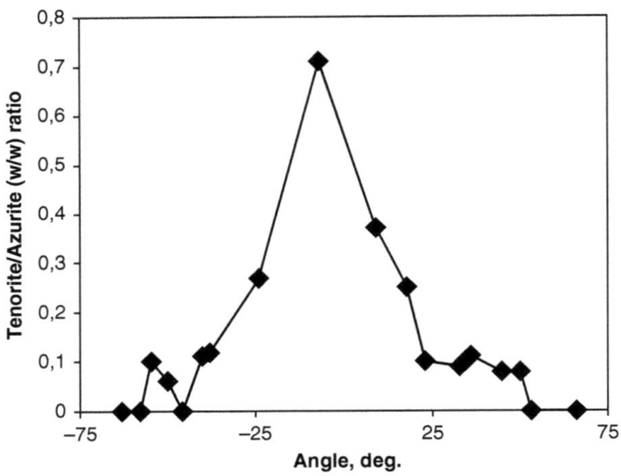

Fig. 4.12 Variation of the tenorite/(azurite + smalt) (w/w) ratio in samples from the *Sant Joan del Mercat* church in Valencia, with the angle from the axis in the vault

4.4 Absolute Quantitation, Standard Addition Method

The main drawback for determining the absolute composition of solid samples using the voltammetry of microparticles approach is that there is no possibility of controlling the exact amount of sample deposited on the electrode surface. To solve this problem, addition of electroactive standards of known concentration can be used in the sample. In the following paragraphs, we describe different methods for determining absolute concentrations in solid materials.

Let us consider a solid sample containing an unknown mass concentration, f, of an electroactive species, A. Assuming, as before, that under the selected electrochemical conditions (technique, parameters, electrolyte) the voltammetric peak current (or peak area), i_A, is proportional to the amount of A transferred to the electrode, one can write

$$i_A = G_A fm/M_A \quad (4.10)$$

where m represents the mass of the A-containing sample deposited on the electrode surface. In a standard additions experiment, a series of spiked samples are prepared by mixing aliquots of the original sample with known amounts of a standard compound containing A and a reference electroactive compound R. In the subsequent voltammetric experiment, a mass m of sample a mass m_A of the standard, which contains a known mass fraction f_A of A, are transferred to the electrode. Then, the measured A-centered signal must satisfy the relationship:

$$i_A = (G_A/M_A)(fm + f_A m_A) \quad (4.11)$$

The deposited sample also contains a mass m_R of a R-containing compound (of molecular mass M_R), in which a mass fraction f_R of R exists. Then, the R-centered voltammetric signal should verify

$$i_R = G_R f_R m_R/M_R \quad (4.12)$$

with G_R the electrochemical coefficient of response for R. If the A-centered and the R-centered peaks are well-separated, and no mutual influence exists, one can separately measure the individual peak currents (or equivalent electrochemical parameter) so that the i_A/i_R ratio becomes

$$i_A/i_R = (G_A/G_R)(M_R/M_A)[fm + f_A m_A]/f_R m_R \quad (4.13)$$

Samples should be prepared by weighting the appropriate amounts of the original sample, the A-containing standard and the R-containing auxiliary compound. Conducting the experiment so that a constant and known m/m_R ratio exists for different samples containing known m_A/m_R ratios, one can write

$$i_A/i_R = (G_A/G_R)(M_R/M_A)[(f/f_R)(m/m_R) + (f_A/f_R)(m_A/m_R)] \quad (4.14)$$

Accordingly, plots of i_A/i_R vs. m_A/m_R should give a straight line of slope $(G_A/G_R)(M_R/M_A)(f_A/f_R)$ and ordinate at the origin $(G_A/G_R)(M_R/M_A)(m/m_R)$

(f/f_R). Then, the absolute mass fraction of A in the sample can be calculated from the abscissa at the origin: $AO = (m/m_R)(f/f_A)$ as [234]:

$$f = (AO)f_A(m_R/m) \tag{4.15}$$

The standard addition methodology can, in general, be applied to any type of voltammetric signal, providing that the aforementioned restricting conditions are valid. A case of particular interest is the use of the stripping methodology previously described. It should be emphasized that the application of this method requires that the signals due to A and R be strictly independent. In the particular case in which oxidative dissolution of metals was used for quantitations, the metals A and R do not form intermetallic compounds, and yield stripping peaks separated by a potential interval of at least 150–200 mV. This situation can be labeled as "electrochemical compatibility" between the metals A and R.

Application of the standard addition method for the determination of the content of zirconium in ceramics is illustrated in Fig. 4.13. Here, ZrO_2 was used as

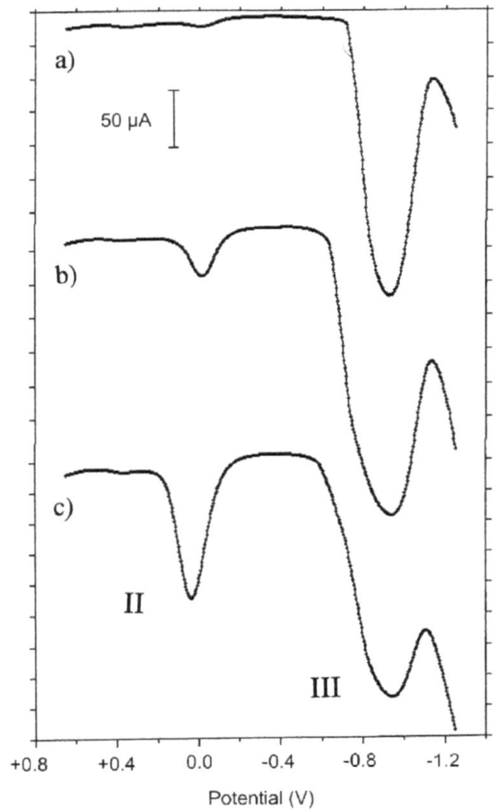

Fig. 4.13 Square-wave voltammograms of PIGEs modified with mixtures of a zirconium-containing sample plus ZrO_2 (standard) and ZnO (auxiliary reference material) in contact with 0.10 M $NaCl$. Sample: ZnO mass ratio equal to 4.422; $ZnO : ZrO_2$ mass ratio equal to (**a**) 0.163; (**b**) 1.216; and (**c**) 5.374. Potential scan initiated at −1.45 V in the positive direction without prior electrodeposition step. Potential step increment 4 mV; square-wave amplitude 25 mV; frequency 15 Hz. [234]

a standard, while ZnO was used as a auxiliary reference material. This compound was selected because of its excellent adherence to graphite electrodes and the electrochemical compatibility between Zr and Zn (which do not form intermetallic compounds). Both compounds are reduced to their corresponding metals at potentials of about −1.25 V. These processes can be represented as

$$ZrO_{2s} + 4H^{+}{}_{aq} + 4e^{-} \rightarrow Zr_s + 2H_2O \tag{4.16}$$

$$ZnO_s + 2H^{+}{}_{aq} + 2e^{-} \rightarrow Zn_s + H_2O \tag{4.17}$$

The square-wave voltammetry of different samples immersed in 0.10 M $NaCl$ yield well-defined anodic stripping peaks at −0.95 and 0.00 V—corresponding to the oxidation of the metallic deposits of Zn and Zr, respectively.

It is important to note that sample dilution with an inert material is, in general, desirable—in particular, when stripping of metals is involved [234]. If large amounts of depolarizer are transferred to the electrode surface, bulky metallic deposits can also be formed, resulting in distorted stripping peaks. This dilution is generally desirable in multicomponent systems to avoid possible intercomponent effects, as described by Grygar and van Oorschot for mixtures of iron and manganese oxides [230]. Silica and clays have been used [232] for this purpose. Figure 4.14 shows typical square-wave voltammograms for mixtures of a Zr-containing sample, plus ZrO_2 (standard) and ZnO (auxiliary reference material), in contact with 0.10 M $NaCl$.

As can be seen in Fig. 4.14a, the metallic deposit obtained after application of a reductive electrodeposition step in ZrO_2 plus ZnO mixtures consists of large irregularly-shaped grains of Zn accompanied by fine Zn column-shaped grains—as indicated by the SEM/EDX analysis depicted in Figs. 4.14b and 4.14c [234].

It should be emphasized that the voltammetric profile depends significantly on the potential of the reductive electrodeposition step and its duration (electrodeposition time). The position of stripping peaks is also sensitive to the presence of complexing agents in the electrolyte solution. It should also be noted that the stripping response in mixed ZrO_2 plus ZnO systems can be conditioned, not only by the possible formation of intermetallic compounds during deposition, but also by the modification of the electrode surface due to the sequential deposition of the metals. Because the deposition of Zr occurs at potentials more negative than those at which the deposition of Zn takes place, the processes of nucleation and growth of Zr metal occur over a partially Zn - plated graphite surface. Accordingly, the deposit of Zr (and consequently, its subsequent stripping dissolution) may differ from that obtained in the absence of Zn deposition. This may result in peak potential shifts, but also in distorted background currents. Under optimized conditions, satisfactory linear-calibration plots were obtained using both peak current and peak area measurements.

Analysis of errors [234] indicated that the optimal conditions for standard addition experiments correspond to values of G_A/G_R and m/m_R of unity. The main limitation of the method is, obviously, the relatively large amounts of sample required (about 1–2 mg) for weighting in conventional analytical balances. This is a

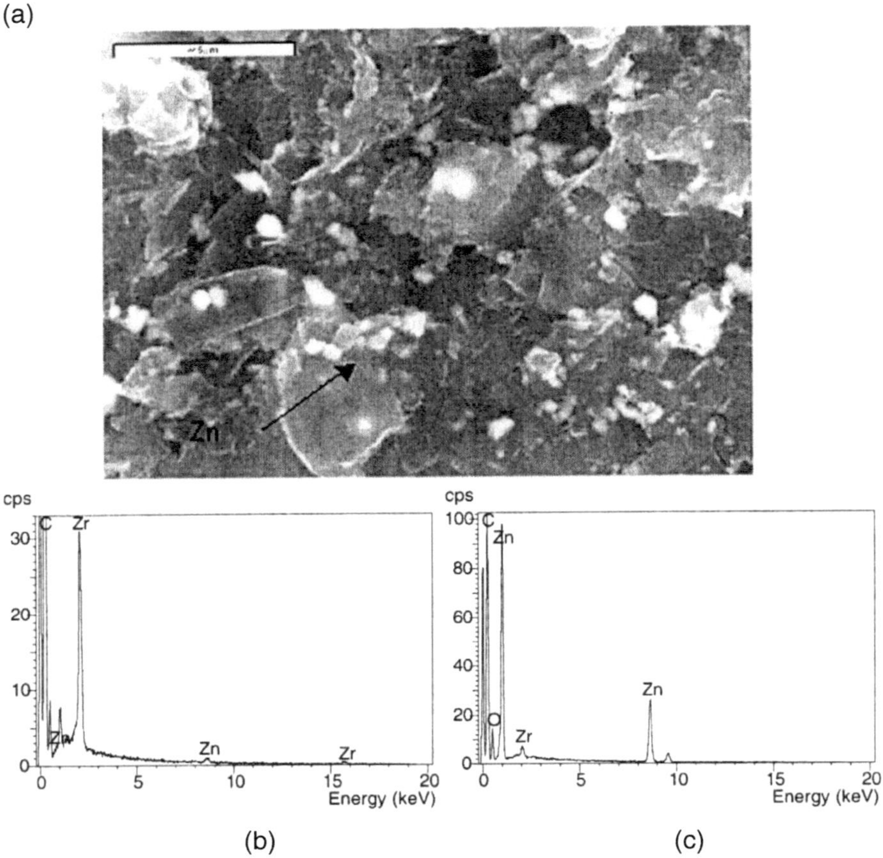

Fig. 4.14 (**a**) SEM image of a deposit of Zr and *Zn* obtained after application of a electrodeposition step at a constant potential of −1.65 V, to a PIGE modified with a 0.50:0.50 (w:w) mixture of ZrO_2 plus *ZnO* immersed into 0.10 M Na*Cl* during 4 h. x-ray energy microanalysis of: (**b**) large irregular grains and (**c**) small column-shaped grains [234]

serious limitation in a number of cases. There are a variety of samples (pristine pigments, ceramics, raw materials, etc.), however, that permit sampling within the required amount.

4.5 H-Point Standard Addition Methods

The above method can be applied for determining the absolute concentration of one or more than one electroactive species in the sample providing that their electrochemical signals are well-separated in the voltammogram and that no mutual influences (interference effect) and matrix effects due to inert, nonelectroactive components in the sample exist.

Frequently, however, the electroactive analytes yield overlapping signals, so that there is no place for applying the "direct" standard addition method previously described. To avoid this problem, a multivariate generalization of the standard addition method was developed by Saxberg and Kowalski for analyzing multicomponent systems [235]. More recently, multivariate methods using partial component and partial least squares have been proposed [236]. These methods suffer from the need of a set of calibration samples with known concentrations of the analyte to be determined. In this context, the H-point standard addition method developed by Bosch and Campins in 1988 allows for a direct correction of proportional and constant errors produced by the matrix in multicomponent samples [237]. This method leads to the determination of the concentration of a given analyte in the presence of a direct interferent, allowing for the determination of the concentration of that interferent. The H-point standard addition method has been applied for determining analytes using a variety of spectrophotometric and chromatographic techniques [238, 239], and has been recently used by Shams et al. for the simultaneous determination of lead and tin in aqueous solution, using differential pulse anodic stripping voltammetry [240].

In the context of the voltammetry of microparticles methodology, the H-point standard addition method has been adapted for determining organic dyes [241] as well as lead and tin in ceramics [242]. Let us consider a mixture of material containing unknown amounts of two electroactive compounds, A and B, and a reference compound, R. It is assumed that weighted amounts of both materials are accurately powdered and thoroughly mixed so that the mass ratio between the A,B-containing material and the reference compound, m/m_R, is known.

If all A, B, and R compounds are electroactive in a suitable electrolyte, voltammograms of the mixture must provide peaks corresponding to their respective redox processes. In solid-state voltammetry, in which separate microparticles of each one of the electroactive compounds are mechanically transferred to the surface of an inert electrode, independent electrochemical processes must occur. Accordingly, peak currents, $i_p(j)(j = A, B, R)$, recorded for A-, B-, and R-centered voltammetric processes, can generally be taken as proportional to the amount of each one of the compounds deposited on the electrode [131, 243–245]. If separate peaks are recorded for A, B, and R, the respective peak currents must satisfy the relationship

$$i_p(j) = G_j r_j m \tag{4.18}$$

where $r_j(j = A, B, R)$ represents the mass fraction of the j-compound in the sample and m represents the mass of that sample transferred to the electrode surface. $G_j(j = A, B, R)$ represents an electrochemical "coefficient of response," characteristic of the redox properties of the analyte (number of electrons transferred, reversibility, diffusion coefficients for electrons and ions), electrochemical conditions (potential scan rate, etc.), and other experimental parameters (electrolyte, pH).

In the most favorable cases, such as in certain stripping processes [79], the current at each potential throughout the voltammetric peak can be taken as proportional to the amount of depolarizer deposited on the electrode. Accordingly, the current measured at a given potential, $i_E(j)$, can be related to the mass fraction of A and B in the sample labeled as r_A, r_B, respectively.

If the voltammetric peaks corresponding to A- and B-localized electron transfer processes are overlapped, the currents at two given potentials, i_1, i_2, in the region in which the overlapped peaks of A and B appear, must satisfy the relationships

$$i_1 = G_{1A} r_A m/M_A + G_{1B} r_B m/M_B \tag{4.19}$$

$$i_2 = G_{2A} r_A m/M_A + G_{2B} r_B m/M_B \tag{4.20}$$

where $G_{jk}(j = 1,2; k = A,B)$ represents a potential-dependent electrochemical coefficient of response—characteristic of the depolarizer and the electrochemical conditions—whereas M_A, M_B again represent the molecular or atomic masses of A and B, respectively.

Because it is not possible to control exactly the amount of sample transferred to the electrode surface, nonrepeatable values of i_1, i_2 will generally be obtained. For this reason, a known amount, m_R, of a third auxiliary reference compound (R) is added so that the sample/reference material mass ratio, m/m_R, is constant and known. If the voltammetric signal corresponding to the reduction or oxidation of R occurs at potentials sufficiently separated from those at which overlapping peaks of A and B are recorded, the peak current measured for R, $i_p(R)$, must satisfy

$$i_p(R) = G_R m_R \tag{4.21}$$

In this equation, G_R represents the corresponding electrochemical coefficient of response.

If the experiment is repeated after adding a known amount of a standard of A, m_A, to the mixture of the sample and the auxiliary reference compound, the currents measured at potentials E_1, E_2 in the overlapping A+B peaks become

$$i_1 = G_{1A}(r_A m + m_A)/M_A + G_{1B} r_B m/M_B \tag{4.22}$$

$$i_2 = G_{2A}(r_A m + m_A)/M_A + G_{2B} r_B m/M_B \tag{4.23}$$

Although the exact amount of the resulting mixtures of solids mechanically attached to the electrode is not known, the quotients m/m_R and m_A/m_R can be accurately determined. Then, the quotients $i_1/i_p(R), i_2/i_p(R)$ can be written as

$$\frac{i_1}{i_p(R)} = \left(\frac{G_{1A}M_R}{G_R M_A}\right)\left[r_A \frac{m}{m_R} + \frac{m_A}{m_R}\right] + \left(\frac{G_{1B}M_R}{G_R M_A}\right) r_B \left(\frac{m}{m_R}\right) \tag{4.24}$$

$$\frac{i_2}{i_p(R)} = \left(\frac{G_{2A}M_R}{G_R M_A}\right)\left[r_A \frac{m}{m_R} + \frac{m_A}{m_R}\right] + \left(\frac{G_{2B}M_R}{G_R M_A}\right) r_B \left(\frac{m}{m_R}\right) \tag{4.25}$$

Equations 4.23 and 4.24 predict that plots of $i_j/i_p(R)(j = 1,2)$vs. m_A/m_R must give straight lines with slopes of $(G_{jA}/G_R)(M_R/M_A)$. Such straight lines intersect at a point characterized by an H-point abscissa, $(m_A/m_R)_H$. In general, the mass ratio of A can be obtained as [241, 242]

$$r_A = \frac{\left(\frac{m_A}{m_R}\right)_H \left(\frac{G_{2A}-G_{1A}}{G_R}\right)\left(\frac{M_R}{M_A}\right) + \left(\frac{m}{m_R}\right) r_B \left(\frac{G_{2B}-G_{1B}}{G_R}\right)\left(\frac{M_R}{M_B}\right)}{\left(\frac{m}{m_R}\right)\left(\frac{G_{1A}-G_{2A}}{G_R}\right)\left(\frac{M_R}{M_A}\right)} \tag{4.26}$$

If potentials 1 and 2 are selected so that the currents of the interferent B are the same (i.e., $G_{1B} = G_{2B}$), the above equation reduces to

$$r_A = -\left(\frac{m_A}{m_R}\right)_H \left(\frac{m}{m_R}\right)^{-1} \tag{4.27}$$

Since the m/m_R ratio is constant and known, the mass fraction of A in the material r_A can be calculated from Eq. 4.26 by inserting the abscissa at the H-point obtained from a standard addition experiment.

The uncertainty associated with the quotient m/m_R can be lowered by weighting relatively high amounts of sample and the reference material. Attributing the same uncertainty, Δm, for mass measurements, the relative uncertainty of the quotient m/m_R becomes minimal for m/m_R equal to unity. The uncertainty in $(m_A/m_R)_H$, however, depends mainly on the uncertainties of m, m_A, m_R and those associated with current measurements. These last ones depend on the uncertainties associated with the measure of the currents in the overlapping wave, and the peak current of the reference redox process that can be related with m, m_A, m_R and the respective electrochemical coefficients of response. In practice, reasonable measurements can be obtained when $0.05 < i_1/i_p(R)$, $i_2/i_p(R) < 20$ whereas the uncertainty in these quantities must be minimal for values close to the unity.

This method has been applied for determining the concentration of alizarin and purpurin—two anthraquinonic dyes only differing by one -OH group in madder pigments. As can be seen in Fig. 4.15, where square-wave voltammograms of (a) alizarin, (b) purpurin, and (c) morin immersed into acetate buffer are depicted, alizarin and purpurin exhibit quite similar voltammetric responses, with peaks at above +0.45 and −0.55 V. Voltammetric peaks for alizarin and purpurin are separated by few milivolts, so that mixtures of both dyes existing in madder pigments show the signals for both components strongly overlapped.

For the application of the H-point standard addition method, morin (a flavonoid dye) is used as an auxiliary standard. Peak resolution is increased by diluting organic dyes with an inorganic matrix. A silica with certified composition was selected on the basis of its good adherence to graphite surfaces and ability to form homogeneous mixtures with powdered dyes. In all three cases (alizarin, purpurin, and morin), voltammograms of binary silica + dye mixtures with dye:silica amounts between 5:95 and 30:70 (w/w) produced sharp peaks at potentials identical to those recorded at dye-modified PIGEs.

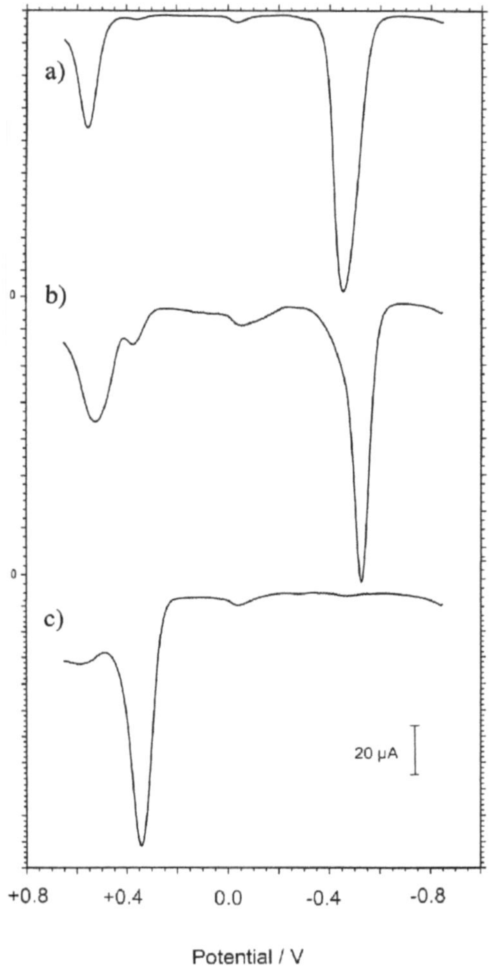

Fig. 4.15 Square-wave voltammograms of (**a**) alizarin, (**b**) purpurin, and (**c**) morin, attached to PIGEs in contact with 0.50 M sodium acetate buffer (pH 4.90). Potential step increment 4 mV; square-wave amplitude 25 mV; frequency 15 Hz [241]

Square-wave voltammograms of ternary silica + morin + alizarin/purpurin mixtures immersed into acetate buffer are shown in Fig. 4.16a, corresponding to a mixture of alizarin plus morin plus silica. Similar results were obtained using phosphate buffer. The response of madder plus morin mixtures is illustrated in Fig. 4.16b, corresponding to a synthetic alizarin plus purpurin plus morin mixture diluted with silica. Here, the morin-centered peak at +0.30 V is accompanied by two pairs of overlapping peaks near +0.55 and −0.45 V. These peaks can unambiguously be attributed to alizarin and purpurin, thus defining a system with highly overlapped peaks.

Calibration curves were obtained for synthetic alizarin plus purpurin samples, using standard additions of either alizarin or purpurin. The dyes were thoroughly

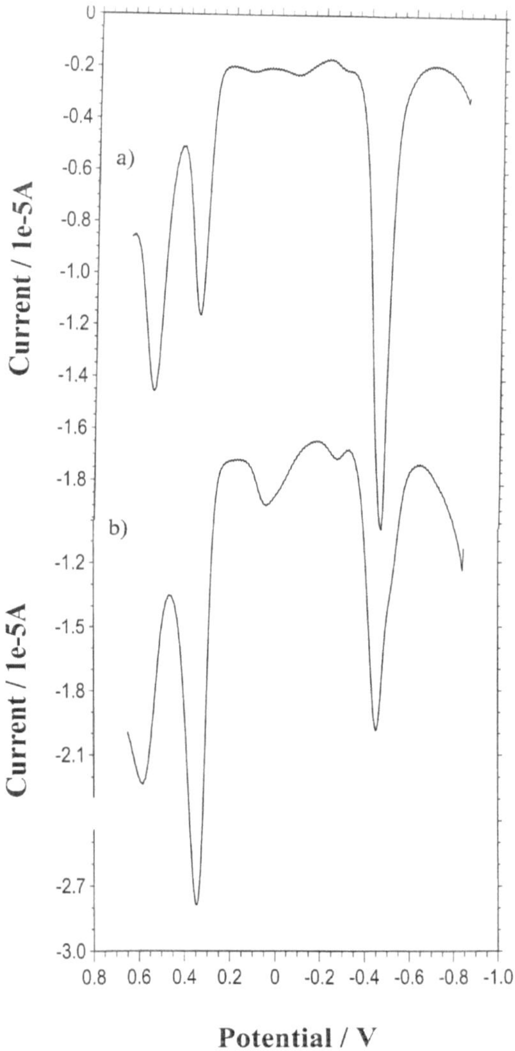

Fig. 4.16 Square-wave voltammograms for (**a**) alizarin (15.72%) + morin (7.16%) + silica (77.12%); and (**b**) alizarin (19.03%) + purpurin (8.30%) + morin (12.48%) + silica mixtures (60.19%). Electrolyte: 0.50 M sodium acetate buffer, pH 4.90. Potential step increment 4 mV; square-wave amplitude 25 mV; frequency 15 Hz [241]

mixed with silica so that the total amount of organic dyes was ca. 10% (w/w). The percentage of morin was between 6% and 9% (w/w). For standard additions of purpurin, currents were measured at −430 and −490 mV in acetate buffer, and at −505 and −620 mV in phosphate buffer—potentials at which identical currents were obtained for the alizarin-centered peak. For standard additions of alizarin, currents were measured at −510 and −555 mV in acetate buffer, and at 575 and −670 mV in phosphate buffer—potentials at which identical currents were obtained for the purpurin-centered peak. As can be seen in Fig. 4.18, well-defined linear H-point plots were obtained.

Table 4.1 Application of the H-point standard addition method to a mixture of a commercial madder pigment (Kremer, 0.03325 g) plus silica (0.25836 g) and morin (0.02486 g), this last acting as an auxiliary reference compound

Additions of Alizarin			$E = -510\,mV^a$	$E = -550\,mV^a$	$E = -575\,mV^b$	$E = -670\,mV^b$
m (g)	m_A(g)	m_A/m_R	$i_1/i_p(R)$	$i_2/i_p(R)$	$i_1/i_p(R)$	$i_2/i_p(R)$
0.01187	–	–	0.436	0.266	0.430	0.270
0.04579	0.00104	0.255	0.542	0.290	0.665	0.310
0.04487	0.00211	0.529	0.655	0.315	0.665	0.310
0.03426	0,00218	0.714	0.731	0.332	0.725	0.335
0.03003	0.00312	1.169	0.920	0.374	0.925	0.380
0.02115	0.00297	1.580	1.090	0.412	1.095	0.410
0.01198	0.00304	2.854	1.618	0.529	1.605	0.535
0.01200	0.00465	4.359	2.240	0.667	2.235	0.665
Additions of Purpurin			$E = -430\,mV^a$	$E = -490\,mV^a$	$E = -505\,mV^b$	$E = -620\,mV^b$
m (g)	m_A (g)	m_A/m_R	$i_1/i_p(R)$	$i_2/i_p(R)$	$i_1/i_p(R)$	$i_2/i_p(R)$
0.01084	–	–	0.418	0.499	0.425	0.500
0.04819	0.00128	0.299	0.477	0.657	0.470	0650
0.04632	0.00252	0.612	0.539	0.822	0.540	0.825
0.03404	0,00262	0.866	0.589	0.957	0.595	0.965
0.02888	0.00260	1.013	0.618	1.034	0.625	1.030
0.01892	0.00275	1.635	0.742	1.363	0.740	1.360
0.01219	0.00274	2.528	0.919	1.835	0.915	1.840
0.01157	0.00462	4.492	1.307	2.871	1.255	2.865

Data from square-wave voltammograms for sample-modified PIGEs in contact with: [a] acetate buffer, pH 4.90, and, [b] phosphate buffer, pH 7.00.

Table 4.2 Statistical parameters for H-point standard addition method applied to commercial madder pigment using silica as a diluent and morin as a reference compound

(a) Additions of Alizarin From Current Data at −510 and −555 mV		
	$i_1/i_p(R)$	$i_2/i_p(R)$
Slope	0.412±0.002	0.0903±0.0018
Ordinate at the origin	0.435±0.004	0.272±0.004
r	0.9999	0.9995
% alizarin in diluted sample	4.9±1.4	
% alizarin in madder sample	43.4±1.3	
(b) Additions of Purpurin From Current Data at −430 and −490 mV		
	$i_1/i_p(R)$	$i_2/i_p(R)$
Slope	0.186±0.003	0.5263±0.0017
Ordinate at the origin	0.429±0.006	0.501±0.003
r	0.9993	0.9997
% alizarin in diluted sample	5.2±1.4	
% alizarin in madder sample	18.1±0.5	

Data from square-wave voltammograms for sample-modified PIGEs in contact with acetate buffer, pH 4.90.

Data in Table 4.1 correspond, respectively, to standard additions of alizarin and purpurin to a mixture of alizarin (5.14%) and purpurin (5.32%) upon addition of morin (4.95%) as an auxiliary reference compound using silica as solid diluent in contact with the acetate buffer. Currents i_1 and i_2 were measured at the above-mentioned potentials, while $i_p(R)$ was measured as the peak current of the morin-centered peak.

Data in Table 4.2 corresponds to the application of the H-point standard addition method to a mixture of a commercial madder pigment diluted with silica, using morin as a reference compound. Calculations were performed by taking $m/m_R = 10.246$, using square-wave voltammetric currents measured for sample-modified PIGEs in contact with an acetate buffer of pH 4.90. Linear plots of $i_1/i_p(R)$ (squares) and $i_2/i_p(R)$ (solid squares) vs. m_A/m_R for additions of purpurin are shown in Fig. 4.17.

Application of this method to the quantitation of lead and tin in archaeological glazes has been reported [242]. Here, SnO_2 and $PbCO_3$ were used as standards for additions, and ZnO was taken as a reference compound. Dilutions were performed with a clay of certified composition free from electroactive species. Typical reductive response was shown in the preceding chapter (Fig. 3.1), consisting of cathodic signals for the reduction of Sn(IV) to Sn(II) and its subsequent reduction to Sn metal and for the reduction of Pb(II) species to Pb metal—all in the 0.0 to −1.00 V potential range. On scanning the potential from −1.25 V in the positive direction, a main oxidation peak corresponding to the oxidative dissolution of Sn and Pb to Sn^{2+} and Pb^{2+} ions in solution is recorded. Overlapping of stripping peaks for tin and lead can be seen in Fig. 4.18, corresponding to (a) SnO_2, and (b) a SnO_2 plus $PbCO_3$ mixture—both diluted with an auxiliary clay. As can be seen in Fig. 4.19, SnO_2

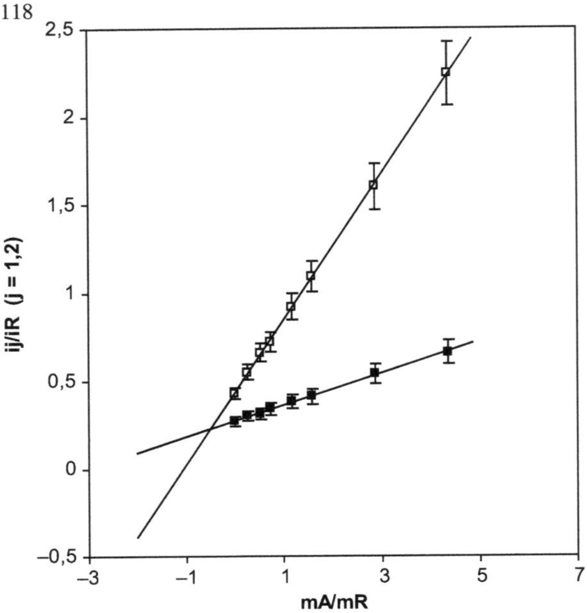

Fig. 4.17 Application of the H-point standard addition method to a commercial madder diluted in silica and using morin as reference compound. Plots of $i_1/i_p(R)$ (*squares*) and $i_2/i_p(R)$ (*solid squares*) vs. m_A/m_R for additions of purpurin. From SQWVs, in 0.50 M acetate buffer, pH 4.90. Potential step increment 4 mV; square-wave amplitude 25 mV; frequency 15 Hz. From data in Table 4.1

plus $PbCO_3$ plus ZnO mixtures present a stripping peak for Zn at −0.80 V, clearly separated from the Sn plus Pb stripping peak at −0.48 V [242].

Results obtained by the application of the H-point standard addition method to three commercial ceramic frits and an archaeological sample from a fragment of ceramic glaze produced in the workshops of Manises (Valencia, Spain) during the first half of the 15th century, are shown in Table 4.3. The calculated values for the percentages of tin and lead are in close agreement with the nominal compositions of the frits and the composition of the archaeological sample obtained by SEM/EDX analysis.

These results illustrate the inherent capabilities of the voltammetry of microparticles for determining the absolute concentration of analytes in samples from works of art. Here, the most serious limitations are associated with (i) the need for well-defined electrochemical responses, and (ii) the need for relatively high amounts of sample. The second limitation, however, does not apply when relative quantitation procedures are used. As a result, a judicious use of such methodologies can provide valuable information for archaeometry, conservation, and restoration.

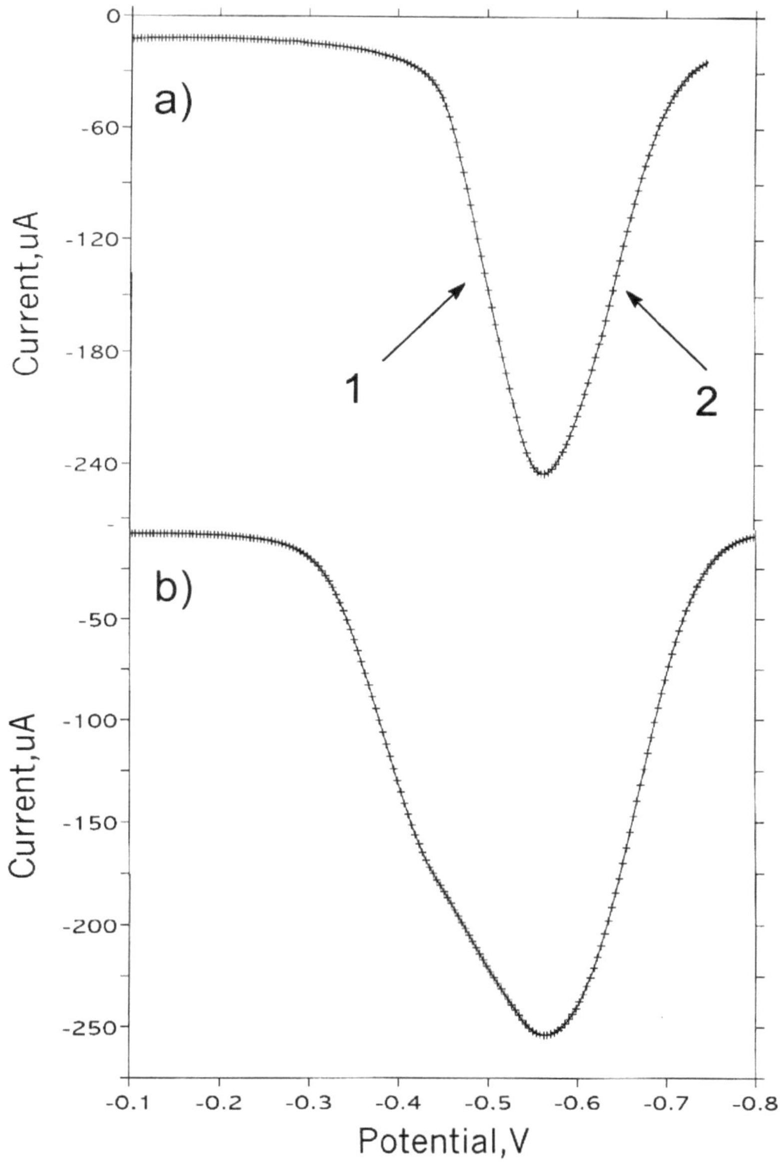

Fig. 4.18 Stripping oxidation peaks recorded for (**a**) SnO_2 plus auxiliary clay (50%, 50% w/w mixture), and (**b**) SnO_2 plus $PbCO_3$ plus auxiliary clay (20%, 40%, 40% w/w mixture) specimens attached to PIGEs in contact with 0.50 M acetate buffer. Square-wave voltammograms initiated at −1.05 V after an electrodeposition step of 30 s at that potential. Potential step increment 4 mV; square-wave amplitude 25 mV; frequency 15 Hz [242]

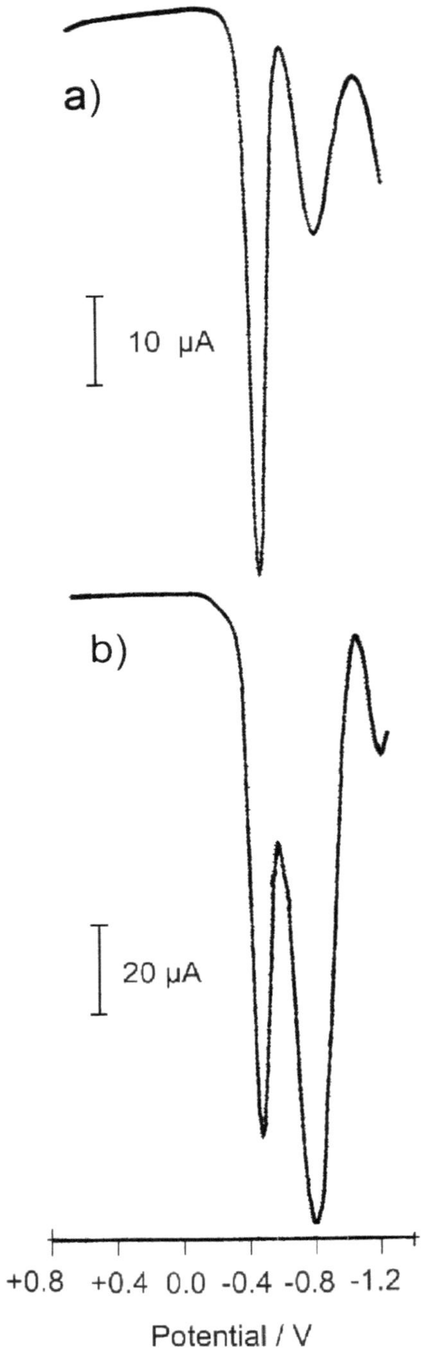

Fig. 4.19 Stripping oxidation peaks recorded for (**a**) SnO_2 plus ZnO plus auxiliary clay (30%, 10%, 60% w/w mixture), and (**b**) SnO_2 plus $PbCO_3$ plus ZnO plus auxiliary clay (10%, 10%, 30%, 50% w/w mixture) specimens attached to PIGEs in contact with 0.50 M acetate buffer. Square-wave voltammograms initiated at −1.25 V after an electrodeposition step of 30 s at that potential. Potential step increment 4 mV; square-wave amplitude 25 mV; frequency 15 Hz [242]

Table 4.3 Comparison of the percentages of Sn and Pb calculated upon application of the H-point standard addition method to three commercial ceramic frits (F-1 to F-3) and an archaeological sample (A-1)—a medieval glazed piece produced in the second half of the 15th century in Manises (Valencia, Spain)

Sample	%Sn (H point)	%Sn (nominal)	%Pb (H point)	%Pb (nominal)
F-1	15.7±0.6	15.84	3.1±0.2	3.25
F-2	8.3±0.3	8.52	5.3±0.8	5.17
F-3	10±1	10.06	42.3±0.3	43.44
F-4	12.0±0.8	11.2	4.9±0.3	4.7

The percentages of Sn and Pb were obtained from electrochemical data using additions of SnO_2 and $PbCO_3$, ZnO as a reference material and 267 silica brick (British Chemical Standards). Electrochemically calculated values from measurements at three pairs of potentials, using the Sn plus Pb stripping peak in acetate buffer (H), are compared with nominal compositions (nom) of the frits and the composition of the ceramic sample obtained from SEM/EDX analysis [242]

Chapter 5
Electrochemical Basis of Corrosion of Cultural Objects

5.1 A Search for Equilibrium

Materials have accompanied mankind virtually from the beginning of its existence. Among all the kinds of materials, and very probably due to their aptitude to form alloys with versatile properties, metals have been of such technological importance that several ancient periods have been named after them: Chalcolithic Period (from Greek: *chalcos* = copper; *lithos* = stone), the Bronze Age, and the Iron Age. This sequence reflects their progressive production, extending from the early use of a few native metals probably some 10000 years ago, and continuing to this day with the development of ever more sophisticated alloys.

Except for the very beginning, when metals were used in their native state (that is, not combined with other elements, as in ores), by hammering to make them harder or heating to make them soft, the production of almost all metals has involved adding energy to the system. The smelting of metals from ores—their separation from oxygen, sulfur and carbon—implies metallurgical processes requiring elevated temperatures, typically in the range of 800–1400°C, and a wide spectrum of redox conditions spanning from highly oxidizing to strongly reducing. As a result of this thermodynamic uphill struggle, the obtained metal has a strong driving force to return to its native, low-energy oxidized state. Such an inevitable comeback (corrosion) is governed by spontaneous reactions with the environment, assuming different forms and processing at rates that depend on many factors. The whole cycle can be represented as in Fig. 5.1 [246].

Whether artifacts are ancient or modern, the principle underlying the corrosion of metallic artifacts is basically the same—an electrochemical reaction occurring between the metal and its environment. Indeed, the paramount factor is the presence of a thin film of moisture on the metallic surface that may arise from rain, mist, or condensation owing to a high relative humidity. Depending on the environmental conditions, such an invisible film may remain on the surface for a while, providing an ideal way for air pollutants to dissolve and reach the metallic surface. The rate and severity of the attack is usually determined by the conductivity of the electrolyte, which depends upon the level of dissolved contaminants. Furthermore, in the case of high relative humidity, dust or grit particles easily adhere to metal

A. Doménech-Carbó et al., *Electrochemical Methods in Archaeometry, Conservation and Restoration*, Monographs in Electrochemistry,
DOI 10.1007/978-3-540-92868-3_5,

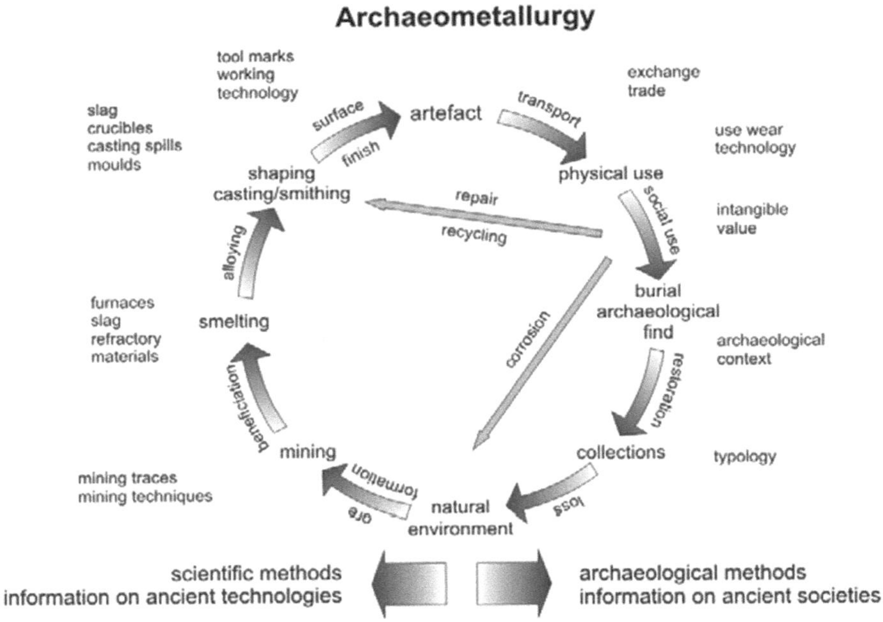

Fig. 5.1 The archaeometallurgy cycle (after Ottaway, adapted by Rehren and Pernicka [1], with permission from the University of Oxford)

surfaces, generating local variations of oxygen partial pressure and hence a potential difference.

Nevertheless, despite similar general rules regulating events occurring at the metal-environment interface, some particularities are evident when approaching the corrosion of metallic antiquities. One of them is the specificity of the object itself, which is essentially chemically and structurally heterogeneous due to ancient production techniques and treatments, as described from old texts that are not always very precise. On the other side, there is the context of where the object is kept, and the environment with which it exchanges continuously. Here there are also some singularities, arising from both the location (burial, museum, outdoors) and the duration of exposure (few days for an exhibition, to centuries and millenaries for archeological finds).

Perhaps the most peculiar factor to be considered when approaching a cultural artifact, however, is its uniqueness. This actuality makes its conservation a challenge—an attractive and interesting case raising numerous difficulties—because the artifact is often the only witness of the past and cannot be sampled to perform all the adequate characterization that would be necessary to understand traditional materials and disclose ancient technologies. Furthermore the alteration compounds (corrosion products) that are considered part of the artifact because they are a testimony of its past, should not be taken away but instead should be studied and conserved in place whenever possible.

5.2 Degradation Under Particular Conditions

The tasks and problems involved in the study of ancient artifacts are manifold, and almost as numerous as the different classes of materials and objects studied by modern specialists. The vast possibilities of interaction between distinct metals and types of environment give rise to unlimited combinations of corrosion forms, making gathering occurrences following traditional classifications useless [247]. Instead of this, and because the exchange with its surroundings is so intensive, metallic heritage can be better understood and conserved when assembled according its context. In this way, three broad categories of artifacts can be delineated—namely archaeological, monumental, and historical artifacts—depending on the environment where they have been (or still are): buried, in the outdoors, or conserved in museums, respectively.

5.2.1 Archaeological Artifacts

Archaeological artifacts are considered to be objects that have been buried for very long periods of time—sometimes many centuries or even millenaries. After such long periods, they are deeply modified, consisting of mostly a mixture of metallic remnants and mineral products, which sometimes hinders their identification (Fig. 5.2).

During the long burial period, extended redistribution of material has taken place. While metal went outwards, ions like chloride and impurities from the environment diffused inwards, resulting in a mass of corrosion products that occupies a volume approximately double the initial size. As a consequence, part of the artifact remains

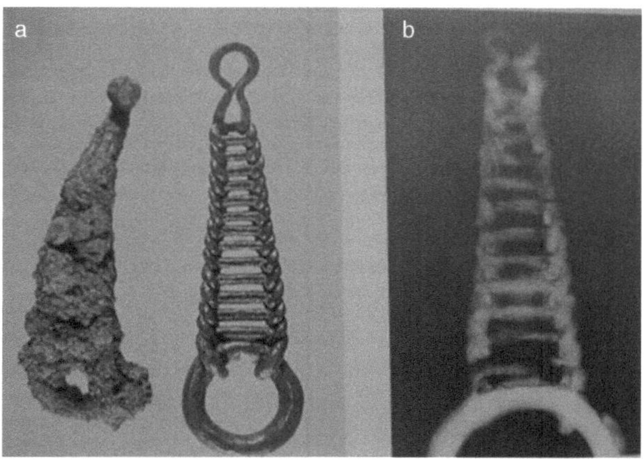

Fig. 5.2 (**a**) The case of a chain: the original shape of the object is modified such that the links cannot be identified anymore; (**b**) radiography of the corroded object is essential to identifying and guiding during cleaning

roughly in place while another part goes outside the original surface. After a while, the build-up of insoluble corrosion products generally stabilizes the artifact even in a corrosive environment. Such an object is said to be "mineralized"—i.e., metal has been continuously transformed into corrosion products, preserving its external morphology as well as signs left by tools used for its manufacture. Due to the importance of identifying and describing the facies, the notion of the limit of the original surface and the surrounding soil has been introduced [248]. This subtle interface can be recognized thanks to special "markers" that exist on each side of it, such as slag inclusions or soil minerals (Fig. 5.3).

Fig. 5.3 Archaeological iron find completely mineralized. Notice the different layers of corrosion products into which it had been transformed over the length of the burial time

When archaeological artifacts are recovered from the ground, the equilibrium state reached during burial may be broken, which is sometimes quite dangerous because the object can become subjected to rapid and irreversible changes unless certain precautions are taken. As a result of a long burial, for instance, iron artifacts often contain soluble salts within the layers of corrosion. Therefore, in ambient air of high relative humidity, water will condensate on the metallic surface and dissolve those salts. Moreover, the formed electrolyte further facilitates the corrosion of the remaining metallic core. This manifests itself as brown droplets on the surface of the object (see Fig. 5.4). The process is so common on excavated iron objects that they often slowly destroy themselves before they have been conserved or fully studied.

Excavated copper alloys are also susceptible to accelerated corrosion even a long time after exhumation due to the presence of chlorides coming from the soil that have attacked the alloy and formed a layer of cuprous chloride (nantokite) deep inside the corrosion layers, just within reach of the metallic core. Cuprous chloride is the stage of a series of reactions in the presence of oxygen and moisture that make it very unstable. The overall gross reaction may be written as [249]:

$$4CuCl_s + O_{2g} + 4H_2O \rightarrow 2Cu_2Cl(OH)_{3s} + 2H^+_{aq} + 2Cl^-_{aq} \quad (5.1)$$

Fig. 5.4 Droplets (see *arrow*) due to the presence of Fe (II) chloride, which progressively oxidizes into Fe(III) and precipitates as oxide and hydroxide

Moreover, the formation of copper trihydroxychlorides accelerates the corrosion of the remaining metal, and is accompanied by an important volume expansion, which results in a fragmentation of the object. This process known as "bronze disease" can be identified by the formation of spots or patches of a light green loose powder on the surface of the object (Fig. 5.5).

Fig. 5.5 Light green powdery eruptions typical of bronze disease

Mechanisms regulating deterioration processes in the burial state are still rather unknown and little research has been done on the consequences of long-term corrosion on the stability of the objects. Nevertheless, some studies have been performed on metal artifacts, seeking to characterize the surface layer as well as establish a relationship between the composition of the corrosion products and the environment where they formed.

The composition of bronze artifact corrosion products excavated from several lacustrine sites could be explained in relation to some characteristics of the environment. In one study, a combination of potential pH diagrams for copper and those defining different types of natural soil was used to explain the composition of corrosion products of copper artifacts excavated from different locations. While sulphides are predominant in anaerobic conditions, mostly carbonates are formed in well-aerated soils [250]. Another investigation performed morphological and chemical characterization of natural patinas on archaeological bronzes from different provenances, pointing out a mechanism of selective dissolution of copper in aerated environments, contrary to the presumed destannification, expected on basis on the relative potentials of both metals [251].

Deterioration mechanisms of iron artifacts have been investigated to bring some insight into long-term corrosion. Measurements using complementary analytical techniques highlight the important role of the transport of species inside the corrosion layer layout—starting at the metal core, a composition gradient is observed going from a dense layer of corrosion products (oxides, oxy-hydroxides, carbonates, and chlorides) through a medium made of a mix of corrosion products and soil compounds to the soil itself at the other extremity [252].

Undersea corrosion was mostly investigated using in-situ techniques, due to the large dimensions of the artifacts that, in addition, are often left in place [253]. Identification of corrosion products on nonferrous artifacts recovered from historic shipwrecks has been performed to gain insight into the long-term degradation process of those metals and alloys, to help conservators determine the most appropriate treatment [254].

By the way, when they are not in an advanced state of mineralization, archaeological artifacts are often structurally fragile. Silver artifacts, for instance, are mostly made from silver-copper alloys and shaped by intensive cold working. Owing to this manufacturing process, the copper-rich phase can be partially caught in the matrix, and gradually released during a long-term burial (precipitation aging). Such precipitates will slowly migrate until they reach and accumulate at the numerous strain lines left in the overworked regions (Fig. 5.6). This region of high concentration of copper-rich phase precipitates and strain lines is mechanically weakened and susceptible to selective corrosion, which leads to fragmentation of the artifact [255].

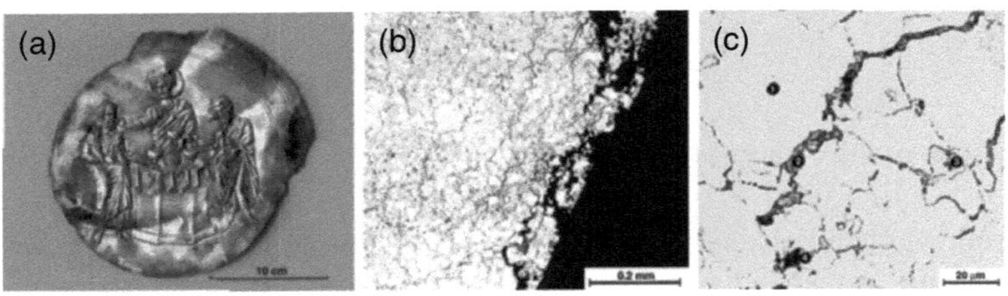

Fig. 5.6 (**a**) Byzantine paten (ca. 600 AD, The Menil Collection, Houston); (**b**) intergranular corrosion; and (**c**) EDX analysis showed primarily silver at location 1 and discontinuous precipitation of copper at locations 2–4; SEM metallographs and analysis courtesy of I. Joosten ICN (after [255])

In any manner, even though some of the classical corrosion forms can be also found in archaeological finds, they are frequently configured in a very complex form—either from a morphological, structural, or chemical viewpoint. Moreover, corrosion products bear a large piece of information about the artifact's "life" and should not be inadvertently removed, contrarily to modern metals, which would simply be etched. It is a big challenge to conserve such objects. On the one hand, it is important to keep most of the mineralized layer because of the information it contains (it is sometimes all that is left from the original object). On the other hand, it is quite necessary to remove harmful contaminants, like chlorides, because they are often confined at the interface between the remnant metal core and the mineralized layers, threatening the stability of the artifact.

5.2.2 *Monuments*

Sculptures and architectural structures and ornaments constitute another category of metallic heritage to be conserved. Their unique character within the broader conservation framework is defined by the considerable range of metals and finishes, sizes, and functions. Moreover, an outdoor location demands treatment that will withstand wind, weather, traffic vibrations, and atmospheric pollution. Most of the objects are also accessible to public touch and vandalism.

Although weathering is an expected evolution corresponding to a natural aging, the increasing air and water pollution following modern industrialization, and the development of transportation in urban areas have had copious consequences on the durability of the objects exposed to the outdoors. In addition to the intrinsic properties of the metallic substrate, the kinetics of the weathering mechanisms strongly depend on the relative humidity and the aggressiveness of the environment. Numerous studies on acid rain, gaseous pollutants (such as SO_2) have quantified the acceleration of the kinetics of damage due to the air pollution and development of new deterioration processes [256]. Nowadays, the concentrations of SO_2 responsible for the greatest damage to outdoor metals overall have decreased, and aerosols and organic acids have become the main agents of material deterioration in outdoor and indoor environments. [257].

Conservation problems in this area usually fall into two basic categories that can be termed either "skin" or "bone"—that is, problems of surface deterioration or problems of structural deterioration, respectively. Surface deterioration regarding copper alloys is by far the most reported case in the literature. Such alloys are generally covered with a more or less stable conversion layer (the patina) mostly originally made for aesthetic purposes [258]. The progressive wearing down of such layers by weathering and pollution inexorably leads to the deterioration of the sculpture. Generally, geodetic lines can be seen on the bottom-faced regions, following the way of rain drain ("zebra striping," see Fig. 5.7a). A close examination reveals that the black areas are more protuberant and correspond to the original surface covered by atmospheric deposits, while the green areas are those corroded and eroded (Fig. 5.7b) [259]. The problem is even worse with gilded sculptures, because the development of local galvanic piles accelerates the formation of corrosion products, which disturb the cohesion between the metallic substrate and the gold foil [260].

Some research on waxes and varnishes have been conducted to protect outdoor sculptures, although there is no universal agreement about their appropriateness

Fig. 5.7 (**a**) Statue belonging to the group "Les bourgeois de Calais" (A. Rodin, cast by A. Rudier, Paris, 1920)—notice on the two arms the vertical lines that follow the rain runoff; (**b**) Detail showing some black zones (squares in relief) that correspond to the original surface covered by the atmospheric deposit, middle in the corroded and eroded surface

[261]. All the other metallic alloys (iron, zinc, lead, etc.) that do not form protective and/or aesthetic patina, are usually covered with paint as protection layer. Anyway, many world treasures have been moved indoors after restoration, as in the case of Marco Aurelio and the Venice Horses.

Beyond the skin of sculpture, there is generally a complex structure intended to fix its different parts together. This structure consists primarily of the armatures, originally made of ferrous alloys to give mechanical stability, but also in an assemblage of different materials—nonferrous joining parts, as well as sand from casting, and other types of fillings. After decades or even centuries of exposure to all weather conditions, these different components are often in quite a bad state of conservation. The formation of a microclimate inside the sculpture leads, for instance, to accelerated corrosion of its ferrous parts, which is even more severe at the points of contact with the more noble cuprous skin (see Fig. 5.8). Moreover, the corrosion products

Fig. 5.8 (**a**) "Le lion de Belfort" (Bartholdi, 1880); (**b**) detail showing a corroded iron armature inside the sculpture

(owing to their bigger volume) generate strains and create an oxidizing environment, which are both very harmful to the monument. To prevent further degradation or for security reasons, the original corroded material has to be partially or completely replaced, although this procedure seems to contradict some deontological principles of conservation. Stainless steel and titanium alloys are nowadays used as substitutes in many instances because they are thought to be highly corrosion-resistant. However, although their corrosion properties are very well known, the application of commercial alloys to the very specific needs of the restoration of historical monuments is far from being well-exploited [262].

Other architectural structures reported in the literature are bridges and roofing, which require different conservation strategies. The Alexandre III bridge in Paris, built at the end of the 19th century, is an example of the summit of the metallic construction of this period. After a century in use, corrosion was evident in regions where water stagnates and particulate matter accumulates, as well in joints and points of contact with cuprous alloys [263] (see Fig. 5.9). To investigate the original color, a cross section from the thick picture layer was made and showed many layers that have been added during previous restoration campaigns.

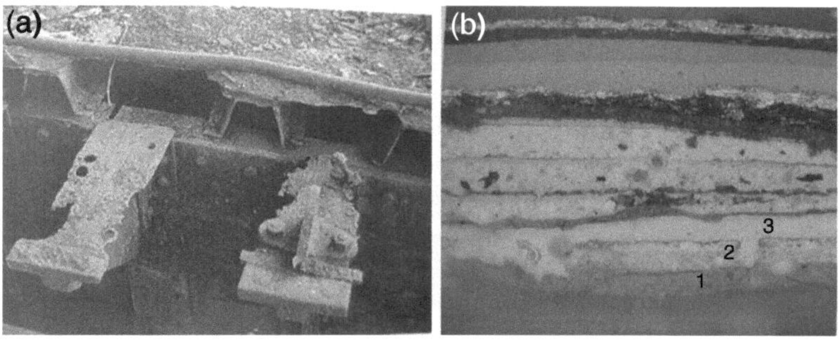

Fig. 5.9 (**a**) Pont Alexandre III, Paris—corroded internal structure; (**b**) cross section evidencing the stratigraphy of paint layers on the Pont Alexandre III, Paris. The three original layers are at the bottom of the image—orange (1), light beige with translucent crystals (2), and light beige (3) (after [263])

5.2.3 *Historic Artifacts*

The third category of metallic artifacts includes collections of a most different provenance—such as scientific instruments, fine arts, historic pieces, ethnographic specimens, etc., which are usually kept in museums. Contrary to the belief that an object is safe once it enters a museum, certain storage or display conditions may lead to corrosive reactions that are different from those found in the natural environment [264, 265]. Some of these dangers come from "off-gassing" from materials used to build display cases and rooms, as well as air pollution introduced by visitors.

The most common gaseous pollutants released by construction materials, are NO_x, SO_2, H_2S, O_3, $HCOOH$, CH_3COOH and $HCHO$, which potentially interact with metal alloys, either directly or as catalytic reagents [266]. The presence of moisture also plays a key role, resulting in the formation of a thin condensation layer on the metallic surface that acts as an electrolyte. This provides a direct way for pollutants to reach the surface, as well as to metallic ions that may move far from it, thus accelerating the corrosion process. Moreover, particulate matter deposited on the object surfaces allows the entrapment of water vapor, which could enhance its deterioration. Depending on the nature of the formed compounds and the tightness of the building, pollutant levels can be high enough to cause strong damage and accelerate the aging of art pieces. However, it is extremely difficult to identify the threshold level of pollutants, which enables damage to the materials because of a possible synergy between several environmental pollutants, such as RH, and the extreme variability of materials housed in collections. Identifying the sources, the nature, and the potential for damage created by indoor pollutants is a prerequisite to minimizing their presence.

By far one of the most reported cases in the literature is the storage of lead artifacts in wooden cabinets. Although very resistant to atmospheric corrosion, lead is extremely susceptible to attack by volatile acids, and the consequences vary from white efflorescence growing from more or less deep pits to the complete disintegration of the specimen into a pile of corrosion products (see Fig. 5.10) [267]. Analysis of the corrosion products points out the presence of lead formiate and/or acetate, which are progressively transformed into basic lead carbonate. Bronzes, on the other hand, are reported to be affected differently according to their composition and also to the nature of pollutants. Volatile compounds emitted by wood and adhesives can attack either the copper-rich matrix, forming green acetates, or the lead-rich regions, where pits progressively spread underneath the surface, provoking an almost complete transformation of the core into corrosion products. Gases containing sulfur are mostly responsible for surface tarnishing.

Silver and silver-copper alloys are often even more sensitive to sulfur compounds—the effect is catalyzed in the presence of oxidizing species and high relative humidity [268]. The resulting surface layer—dark, thin and adherent when developed on a polished surface—usually acts as a good physical barrier that slows down the tarnishing reactions. However, since silver objects are expected to be shiny,

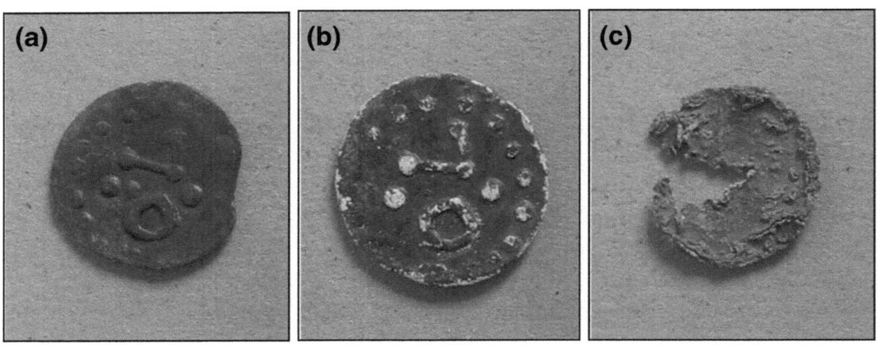

Fig. 5.10 State of lead coins after different conservation conditions: (**a**) coin protected by a passive surface; (**b**) coin with corrosion localized on protruding features; (**c**) coin nearly disintegrated due to corrosive attack (courtesy Société Archéologique de Namur, from [267])

the repetition of the tarnish-polish cycle leads to important losses [269]. Moreover, products generally used to conserve the silver shine can have some harmful side effects. On one hand, cleaning mixtures are extremely difficult to rinse away because they usually stick in the hollow parts of the object where humidity and dirt tend to accumulate. On the other hand, the application of a protective lacquer, if not evenly done, will leave missed parts that may corrode at an enhanced rate [270].

The few reported cases concerning other metals, like zinc, aluminum, and magnesium, attest their susceptibility to corrosion due to volatile compounds in the museum environment [271]. Iron is naturally vulnerable to atmospheric corrosion whatever the pollutants, and the conservation of ferrous artifacts implicates a precise control of relative humidity, often requiring a surface protection like varnish, wax, or oil [272].

For all the above reasons, the concern about environment characterization and monitoring is expanding and a new discipline—preventive conservation has emerged. Its goal is to adapt the environment to materials constituting artefacts, thus minimizing further deterioration.

5.3 Some Useful Corrosion

"Like the electrolytic couple that underlies it, corrosion has two sides, and a glance at history shows that much useful techniques have been stimulated by it." [273] Evidence of the early use of controlled corrosion is provided by the ancient literature. For instance, ceruse (basic lead carbonate) and verdigris (basic copper acetate) have been obtained for the making of pigments by corroding lead and copper, respectively, while iron was corroded to yield crocus powder (powdered iron oxide used for polishing), or put into solution to make ink, or etched with a copper-bearing solution to obtain a rough copper-plate surface as a basis for amalgam gilding [274].

The electrochemical replacement of one metal by another was noticed by Pliny in antiquity. He mentioned that iron looks like copper if it is smeared with "vinegar or alum" previously ground in a bronze mortar [275]. Metal replacement seems to have also been used in the Roman era for tin-plating bronze, as well as underlying the recipes for the preparation of iron surfaces to receive amalgam gilding that are given in the 9th century [276].

Another interesting example, used in different times and in various regions around the world, is the selective removal of copper from its alloys with the noble metals. In South America, most pre-Columbian "gold" objects are actually made of alloys that contain less than 30% of the precious metal. A pure gold surface was nevertheless obtained after treatment with corrosive natural minerals mixed with salt. Dealloying led to the formation of a microporous surface layer of pure gold, which was further consolidated by burnishing to close the pores (see Fig. 5.11). This "depletion gilding" process has been identified by studying the alloy composition beneath the surface of the object [277]. A similar result was obtained with silver-copper alloys, where copper is first oxidized to form an external layer of scale and

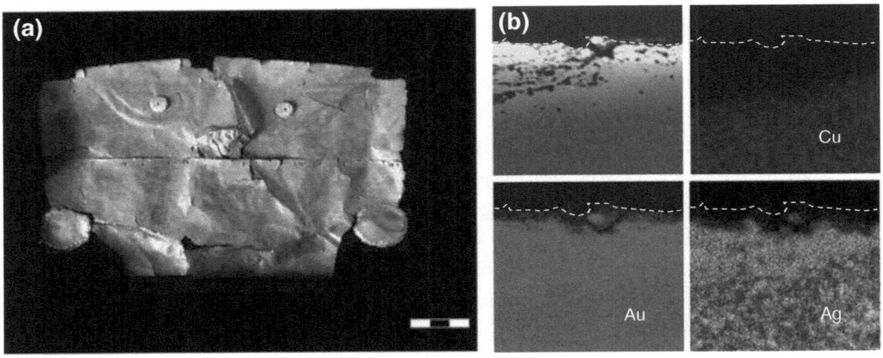

Fig. 5.11 (**a**) Pre-Columbian mask made of ternary alloy (ca. 43% Au, 35% Ag, 21% Cu), courtesy Museu de Arqueologia e Etnografia, USP, Brazil; (**b**) Cross section SEM image and corresponding EDX mapping. The dashed line is the object's contour. Note the strong Cu depletion at distances up to 10–15 μm from the surface

subsequently removed by pickling. The resulting surface is strongly enriched in silver (see Fig. 5.12) [278]. Such a process, known as blanching, was often used by mints to let the coins appear as if they were of higher value.

The knowledge and control of localized corrosion was also required to produce etched plates for the graphic artist with its intaglio lines. Prints made by the process first appear around 1500 AD, but there was a long prehistory in the decorative etching of metal surfaces, particularly those of steel arms and armor [279].

Last, but not least, there is the intentional coloring or patination of metals, of prime importance in sculpture and decorative arts, as this is (in many cases) the key factor in the visual coherence and significance of objects. There is some evidence of their early use in Chinese bronze mirrors, in Japanese sword guards made with alloys especially developed for patination, in the classical Greek and Roman bronzes, in Islamic metalwork, and in cast bronze sculptures from the Renaissance to the present day [280].

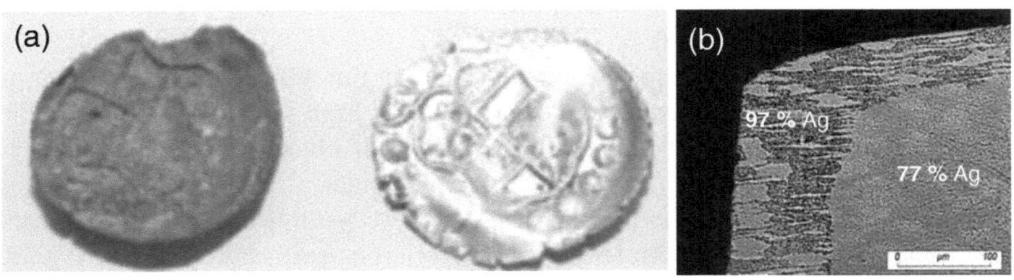

Fig. 5.12 (**a**) Medieval coins from the Preuschdorf's hoard, as found (*left*) and after cleaning with chemicals (*right*); (**b**) Cross section of a test sample (80% Ag/20% Cu) showing the influence of 48 h of immersion in a cleaning solution ("silver dip") on the surface composition (after [278])

Chapter 6
Electrochemistry in Treatment and Conservation of Metal Artifacts

6.1 Electrochemical Treatment of Metal Artifacts

6.1.1 Historical Evolution

Electrochemistry has been used for more than a century in the treatment of ancient metal artifacts [281]. Ideally, this technique should be able to reverse the corrosion processes that have progressively transformed the metal into an ionic compound. Depending on the conservation state of the artifact, priorities have to be attributed and the treatment will be different if consolidation, stabilization, or cleaning is privileged.

6.1.2 Cleaning

A cleaning treatment used to be applied to artifacts with a good metallic structure, whose surface is generally covered with a thin layer consisting in a mixture of corrosion products and grime, sometimes called "tarnish." Cleaning aims to remove this undesired superficial layer, without (or with minimal) loss of the metallic substrate. In many cases, such a goal is more easily achieved by electrochemical methods than by mechanical and chemical methods [282].

The "galvanic" or "contact" method was the precursor to electrochemical treatment [283]. In an electrolytically conducting solution, the artifact is brought into contact with a piece of a less noble metal—usually zinc or aluminium. While the metal corrodes, the electrons supplied to the object allow the reduction of the tarnish layer. Although simple, this method presents some drawbacks, like the progressive contamination of the solution by the corrosion products of the active metal and also limitations with respect to the choice of applicable solution.

Electrochemical cleaning using an external power supply allows the use of a chemically inert anode—ideally platinum, but usually stainless steel or graphite. The solution composition and pH may be adapted at will, depending on the nature and solubility of the compounds that must be reduced. Moreover, the rate of the reduction reaction can actually be controlled. Most reports on application of this

A. Doménech-Carbó et al., *Electrochemical Methods in Archaeometry, Conservation and Restoration*, Monographs in Electrochemistry,
DOI 10.1007/978-3-540-92868-3_6, © Springer-Verlag Berlin Heidelberg 2009

technique concern silver objects using diverse electrolytes such as sodium carbonate [284] and formic acid [285, 286]. Iron objects have been cleaned in sodium hydroxide [287]. Lead tokens were treated in diluted solutions of sodium hydroxide or sulphuric acid [288].

A power supply is also often used when the goal is not just transforming but also removing a thick layer deposited on the surface as a consequence of a long burial period, as in the case of cannons submerged in seawater. In this case, a very negative potential is applied to promoting strong hydrogen evolution that generates gas bubbles that are able to blow off the encrustation and mechanically clean the object [289].

Finally, electrochemical cleaning using a potentiostat enables the application of a well-defined potential to the cathode (artifact) in order to select the reaction that must take place. In addition, monitoring the current gives information about the completion of the reduction of a specific corrosion product. Proposed initially for the treatment of lead tokens [290], this selective technique has especially been used to clean thin metallic coatings on metals, like silver on cuprous alloys [291], gold on silver [292], or other materials like silvered flag fringes [293]. The treatment could also be performed locally, with the help of a "pencil" [294].

6.1.3 Stabilization

One important application of electrolytic treatment is the removal of harmful anions, such as chloride and sulphide, from the mineralized archaeological artifacts. The negative polarization of the system repels the negatively charged species out of the cathode. The process is often accompanied by the formation of either gas or soluble species in the electrolyte. This kind of treatment was carried out to increase the rate of extraction of chlorides from iron (see Fig. 6.1) [295], copper [296], and aluminium [297] mineralized objects.

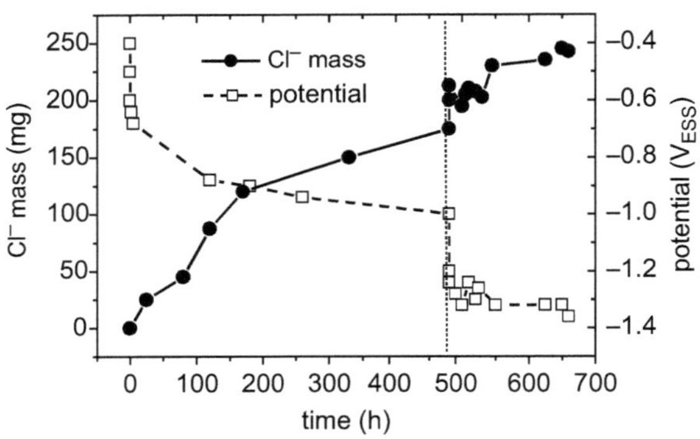

Fig. 6.1 Curve showing the chloride egress (*filled squares*, left axis) from the artifact (dosed in solution), and corresponding variations of the rest potential (*open squares*, right axis) of the object. After 500 h, a potential $U = -1.52$ V was applied, which accelerated the rate of removal of Cl^- (after [295])

6.1.4 Consolidation

Consolidative reduction has been employed when the object is so badly corroded that it becomes extremely fragile and all surface details are just a mass of corrosion products. In this case, one must apply a low-current density over a prolonged period. In some instances, it is necessary to hold together the loose crusts during the electrolysis, which are then ultimately reduced in situ into a more coherent but porous metallic network. In the case of a completely mineralized silver lyre from the Royal Graves at Ur, partially-rectified current was used to reduce silver chloride [298].

More recently, treatments conducted with a potentiostat offer the advantage of maintaining the applied potential at the desired value, which is an important parameter when the treatment is long. Lead communion tokens were polarized in a dilute sulphuric acid solution at a potential just before the onset of the hydrogen evolution. As a result, light grey carbonate encrustations took on a dull grey metallic appearance, and the time of treatment was shortened [290]. Progressive reduction of the corrosion layer during potentiostatic treatment of lead objects was indicated by cross-section examination [299], and the performance of the potentiostatic arrangement to reduce silver chloride (horn silver) has been compared with other reduction techniques used in the treatment of brittle silver [300].

6.1.5 State of the Art

At present, electrochemical treatments are being used as routine in a few workshops in different countries. Most applications deal with detachment of concretions and stabilization of archaeological finds. On a reduced scale, the cleaning of tarnished layers on silver artifacts is also obtained by electrochemical treatment. Besides regular use, little research has been conducted to better understand reactions taking place during the treatment, which should help optimizing operational conditions and evaluating the consequence on the long-term conservation of the artifact.

Applied research recently reported in the literature has been conducted mostly on marine iron wrecks, aiming at evaluating the efficiency of chloride removal from the artifacts from a marine site under cathodic polarization [301] as well as the use of impressed current-cathodic protection to conserve parts of the first ironclad warship [302].

Regarding more fundamental research, two interesting studies have been conducted lately. One of them used different techniques to investigate the electrolytic reduction of corroded lead objects [303]. Observation of cross sections during the reduction process proved that the reactions start at the metal/corrosion interface and evolve progressively outwards (see Fig. 6.2). The resulting morphological and chemical changes were also investigated, confirming a complete reduction of the products present on the surface—hydrocerussite ($Pb_3(CO_3)_2(OH)_2$) and plumbonacrite ($Pb_{10}(CO_3)_6O(OH)_6$)—into metallic lead before the treatment. Extending this phenomenon to some lead alloys usually employed in artifacts, a similar treatment proved to be efficient for the reduction of the Pb-Sb alloy, while no reduction of corrosion products was observed for Pb-Sn and Pb-Bi.

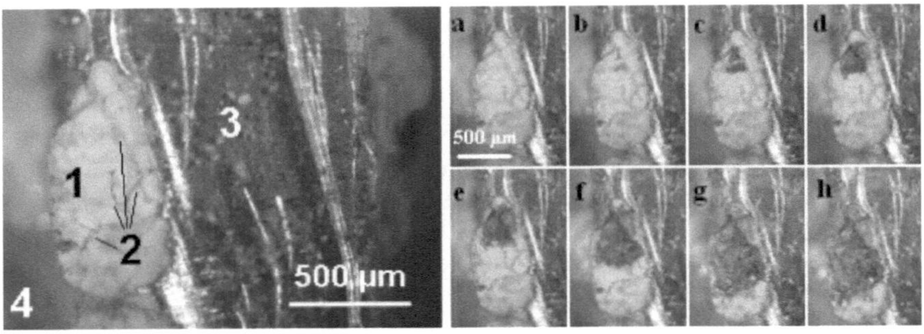

Fig. 6.2 Cross section of a corroded lead artifact before reduction (left): the different regions are (1) the porous corrosion patch, (2) some metal veins, (3) the metal core and (4) the electrolyte. Time-elapsed optical images (**a–h**) of the same section (right) during the reduction of the corrosion layer (from [303])

The second study was designed to contribute to the current focus on the protection of silver artifacts by electrochemically grown poly-3-amino1,2,4-triazole (PAta) [304]. There is evidence that the dissolution of a silver copper alloy is inhibited by Ata adsorption with interaction with oxidation species (see Fig. 6.3).

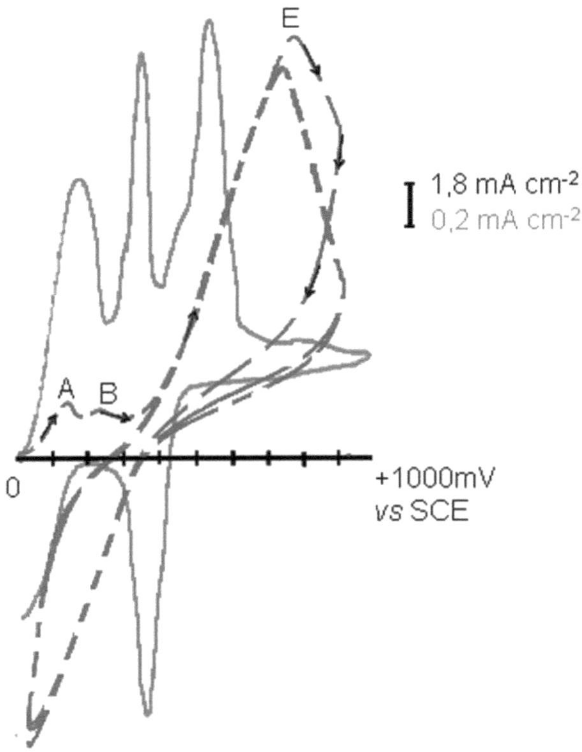

Fig. 6.3 Cyclic voltammogram (1st cycle) of sterling silver (92.5% Ag, 7.5% Cu) in 0.5M NaCH3COO, pH 4, v = 10 mV/s; copper and silver oxidation peaks, indicating surface passivation in presence of Ata (3-amino,1,2,4-triazole) (from [304])

6.2 Evaluation of Museum Environment

Sources of potentially dangerous environmental conditions are currently being investigated with the goal of minimizing the deterioration of cultural artifacts. Because pollutants usually occur at very low concentration levels inside cultural institutions, their determination is a difficult task requiring sensitive, relatively expensive, and not always accessible techniques. Moreover, even though the methods are able to detect few ppb (parts per billion) of a given pollutant, this isolated result can hardly be used to define the level of an environment's harmfulness, because they do not take in account the synergetic effects that may exist when several compounds combine to interact with the surface of a material.

Among the methods used to study the impact of the environment on the deterioration of metallic objects, electrochemical evaluation of metallic passive monitors previously exposed to a given environment appears to be a very suitable tool [305]. Metallic surfaces are indeed highly sensitive to their surroundings, and prone to reacting with airborne pollutants to form compounds that can then be identified by electrochemical reduction. The technique allows for probing of the totality of the layer and its individual chemical constituents. Moreover, the high sensitivity of the technique allows the detection of extremely thin surface films, well before they become visible, providing a good estimation of the reaction rate.

This technique has been recently applied in different studies. One of them evaluated the environmental conditions in the treasure rooms of Reims Cathedral, where silver sensors were exposed during a five-year period [306]. The reduction curves point out that the results strongly depend on whether the coupons were displayed inside or outside a case. Regarding the nature of the products developed on the surface during the exposure, only silver sulphide has been found on the coupons kept inside the showcase, while the additional peak found with coupons exposed outside the case suggests the additional presence of silver chloride (see Fig. 6.4a). Moreover,

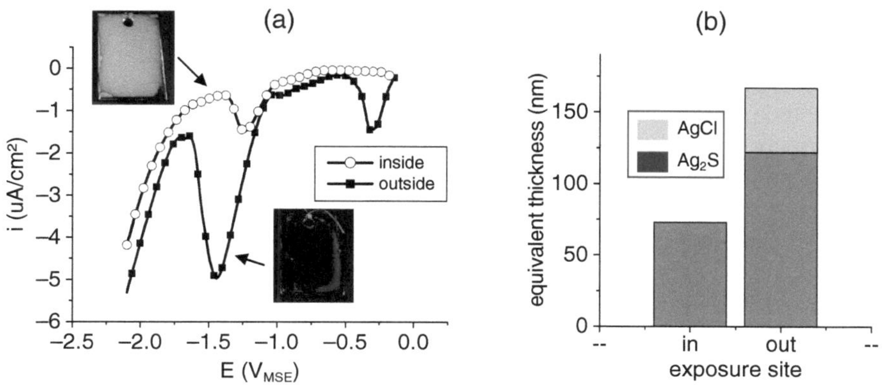

Fig. 6.4 (**a**) Surface aspect and corresponding LSV of coupons exposed for five years inside and outside a showcase (Treasure room, Palais de Tau, Reims). The peaks beginning at −1 and −0.1 V_{MSE} correspond to Ag_2S and AgCl, respectively. Electrolyte: 0.1 M sodium acetate, v = 5 mV/s. (**b**) Equivalent thickness of reduced compounds in the two locations (note the different partition of the two compounds)

the amount of surface product that has been developed in each case, calculated by integrating the area under the current peaks, appears to be two times larger for the coupons exposed outside (see Fig. 6.4b). These results attest to the positive "screening" effect provided by the showcases, even though they are old and not tight.

Attempting to evaluate the effect of particulate matter on the tarnishing of sterling silver more precisely, a specially designed holder was used so that air dust and particles will deposit only on one side of the coupons. The electrochemical reduction of the tarnish layers formed during short-time exposure was performed simultaneously on both sides, using a double compartment cell. Results showed a larger tarnishing rate on the upper side of coupons—i.e., the one exposed to particle deposition. This seemed to be attributed to the selective attack of a copper-rich phase (see Fig. 6.5) [307, 308].

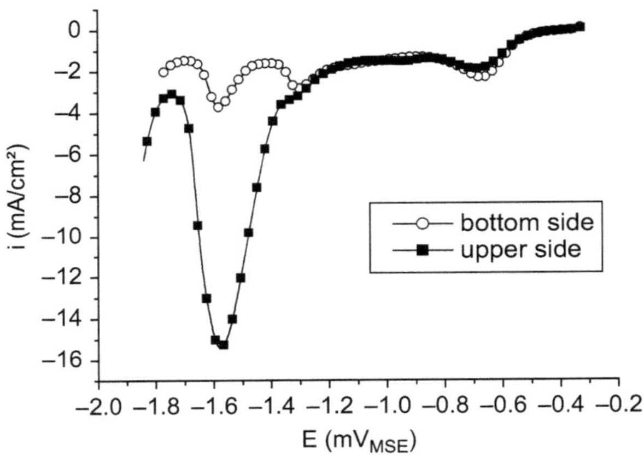

Fig. 6.5 Typical negative-going LSV recorded on the upper (*black squares*) and bottom (*open circles*) face of a sterling silver coupon exposed in cultural institutions. The larger peak intensity measuring at −1.6 on upper face indicates a selective attack of this face that is probably due to the deposition of particulate matter

In subsequent studies, this first approach was extended by also exposing lead coupons, for they are very sensitive to organic acids. After only a few months of exposure, the surface products could be characterized by performing Raman analysis combined with electrochemical reduction. Results point to the presence of formiates and acetates for the coupons presenting the larger corrosion rates [309].

Bibliography

1. Reedy TJ, Reedy ChL (1988) Statistical Analysis in Art Conservation Research. The Getty Conservation Institute, Los Angeles.
2. Lahanier et al. Cited in Janssens K (2005) A survey of the recent use of X-ray beam methods for non-destructive investigations in the cultural heritage sector, Cultural Heritage Conservation and Environment Impact Assessment by Non-Destructive Testing and Micro-analysis, Van Grieken R, Janssens K (Eds) AA Balkema Publishers, London, pp. 265–308.
3. Ramos PM, Ferré J, Ruisánchez I, Andrikopoulos KS (2004) Fuzzy logic for identifying pigments studied by Raman spectroscopy. Appl Spectrosc 58: 848–854
4. Nadolny J (2003) The first century of published scientific analyses of the materials of historical painting and polychromy, *circa* 1780–1880. Rev Conserv 4: 39–51 and references therein.
5. Lahanier C (1991) Scientific methods applied to the study of art objects Mickrochim. Acta 2: 245–254.
6. Velios A, Harrison J (2007) Digital reconstruction of fragmented archaeological objects. Stud Conserv 52: 19–36.
7. Nevin A, Comelli D, Valentini G, Anglos D, Burnstock A, Cather S, Cubeddu R (2007) Time-resolved fluorescence spectroscopy and imaging of proteinaceous binders used in paintings. Anal Bioanal Chem 388: 1897–1905.
8. Asmus FJ (2003) Non-divestment laser applications in art conservation. J Cult Herit 4: 289–293.
9. Saunders D, Billinge R, Cupitt J, Atkinson N, Liang H (2006) A new camera for high-resolution infrared imaging of works of art. Stud Conserv 51: 277–290.
10. Fischer C, Kakoulli I (2006) Multispectral and hyperspectral imaging technologies in conservation: current research and potential applications. Rew Conserv 7: 3–16 and references therein.
11. Gómez ML (1998) La restauración. Examen científico aplicado a la conservación de obras de arte. Cátedra, Madrid.
12. Römich H, López E, Mees F, Jacobs P, Cornelis E, Van Dyck D, Doménech-Carbó MT (2005) Microfocus X-ray computed tomography (mCT) for archaeological glasses. In Cultural Heritage Conservation and Environment Impact Assessment by Non-Destructive Testing and Micro-analysis, Van Grieken R, Janssens K (Eds), AA Balkema Publishers, London, 37–48.
13. Camaiti M, Casieri C, De Luca F, Fantazzini P, Terenzi C (2007) The use of portable single-sided relaxometry and laboratory imaging NMR devices in stone conservation. Stud Conserv 52: 37–49.
14. Jakiela S, Bratasz L, Kozlowski R (2007) Acoustic emission for tracing the evolution of damage in wooden objects. Stud Conserv 52: 101–109.
15. Castellini P, Esposito E, Marchetti B, Paone N, Tomasini EP (2003) New applications of scanning laser doppler vibrometry (SLDV) to non-destructive diagnostics of artworks: mosaics, ceramics, inlaid wood and easel painting. J Cult Herit 4: 321s–329s.

16. Kulicki J (1991) Use of thermography as an investigatory method in conservation research: outline of problems in science, technology, and European cultural heritage: proceedings of the European symposium, Bologna, 13–16 June, Butterworth-Heinemann Publishers, 566–570.
17. Mairinger F, Schreiner M (1982) New Methods of Chemical Analysis – A Tool for the Conservator, Science and Technology in the Service of Conservation, IIC, London, 5–13
18. Mairinger F, Schreiner M (1986) Analysis of supports, grounds and pigments, PACT 13, Xth Anniversary Meeting of PACT Group, van Schoute R, Verougstracte-Marcq H (Eds), Louvain-la Neuve, 171–183, and references therein.
19. Bitossi G, Giorgi R, Mauro M, Salvadori B, Dei L (2005) Spectroscopic techniques in cultural heritage conservation: a survey. Appl Spectrosc Rev 40: 187–228
20. Janssens K, Vittiglio G, Deraedt I, Aerts A, Vekemans B, Vincze L, Wei F, Deryck I, Schalm O, Adams F, Rindby A, Knöchel A, Simionovici A, Snigirev A (2000) Use of microscopic XRF for non-destructive analysis in art and archaeometry. X-ray Spectrom 29: 73–91.
21. Mantler M, Schreiner M (2000) X-ray fluorescence spectrometry in art and archaeology. hboxX-Ray Spectrom 29: 3–17.
22. Leute U (1990) Spectroscopy and archaeometry. Spectrochim Acta Rew 13: 167–190.
23. Ravaglioli A, Krajewski A (1989) Chimica Fisica Tecnica e Scienza dei Materiali Antichi Ceramici e Vetrosi, Museo Internazionale delle ceramiche, Faenza, pp. 260–268.
24. Tykot RH, Young SMM (1996) Archaeological applications of inductively coupled plasma-mass spectrometry, In Archaeological Chemistry: Organic, Inorganic and Biochemical Analysis, Orna MV (Ed) ACS Symposium Series 625, 116–130.
25. Wang Q (2007) An investigation of deterioration of archaeological iron. Stud Conserv 52: 125–134.
26. Yellin J (1995) Neutron activation analysis: impact on the archaeology of the Holy Land. Trends Anal Chem 14: 37–44.
27. Demortier G (1989) Accelerator-based spectroscopic techniques for analysis of archaeological gold jewelry. Spectrosc Int 1: 36–43.
28. Koob SP, Ng WY (2000) The desalination of ceramics using a semi-automated continuous washing station. Stud Conserv 45: 265–273.
29. Niven B, Campbell L (2006) Effects of selected aqueous treatments on the properties of two papers. Stud Conserv 51: 189–198.
30. Wolbers R, Landrey G (1987) The use of direct reactive fluorescent dyes for the characterization of binding media in cross sectional examinations. Preprints of 15th Annual Meeting of the American Institute for Conservation and Artistic Works, Vancouver, 168–202.
31. Heginbotham A, Millay V, Quick M (2006) The use of immunofluorescence microscopy and enzyme-linked immunosorbent assay as complementary techniques for protein identification in artist's materials. J Am Inst Conserv 45: 89–105.
32. Ashok R (Ed.) (1993) Artists' Pigments. A Handbook of their History and Characteristics, vol 2, National Gallery of Art, Washington.
33. Billmeyer FW Jr, Kumar R, Saltzman M (1981) Identification of organic colorants in art objects by solution spectrophotometry: pigments. J Chem Educ 58: 307–313.
34. Roqué J, Molera J, Sciau P, Pantos E, Vendrell-Saz M (2006) Copper and silver nanocrystals in lustre lead glazes: development and optical properties. J European Ceramic Society 26: 3813–3824.
35. Nevin A, Cather S, Anglos D, Fotakis C (2006) Analysis of protein-based binding media found in paintings using laser induced fluorescence spectroscopy. Anal Chim Acta 573–574: 341–346.
36. Casadio F, Toniolo L (2001) The analysis of polychrome works of art: 40 years of infrared spectroscopic investigations. J Cult Heritage 2: 71–78.
37. Derrick MR, Stulik D, Landry MJ (1999) Infrared Spectroscopy in Conservation Science, Getty Conservation Institute, Los Angeles.
38. Edwards HGM, Chalmers JM (Eds) (2005) Raman Spectroscopy in Archaeology and Art History, The Royal Society of Chemistry, Cambridge.

39. Vandenabeele P, Edwards HGM, Moens L (2007) A decade of Raman spectroscopy in art and archaeology. Chem Rev 107: 675–686.
40. Ghisalberti EL, Godofrey IM (1998) Application of nuclear magnetic resonance spectroscopy to the analysis of organic archaeological materials. Stud Conserv 43: 215–230, and references therein.
41. Spyros A, Anglos D (2006) Studies of organic paint binders by NMR spectroscopy. Appl Phys A 83: 705–708.
42. Dietemann P, Kälin M, Zumbühl S, Knochenmuss R, Wülfert S, Zenobi R (2001) A mass spectrometry and electron paramagnetic resonance study of photochemical and thermal aging of triterpenoid varnishes. Anal Chem 73: 2087–2096.
43. Peris-Vicente J, Garrido-Medina R, Simó-Alfonso E, Gimeno-Adelantado JV, Doménech-Carbó MT (2007) Infusion mass spectrometry as a fingerprint to characterize varnishes in oil pictorial artworks. Rapid Commun Mass Spectrom 21: 851–856.
44. Ballistreri A, Fichera M, Fifi AP, Musumarra G (1998) Characterisation of Vasari and Zuccari binding media by direct pyrolysis in the mass spectrometer. Sci Tech Cult Heritage 7: 27–30.
45. Languri GM, Boon JJ (2005) Between myth and reality: mummy pigment from the Hafkenscheid collection. Stud Conserv 50: 161–178.
46. Scalarone D, Duursma MC, Boon JJ, Chiantore O (2005) MALDI-TOF mass spectrometry on cellulosic surfaces of fresh and photo-aged di- and triterpenoid varnish resins. J Mass Spectrom 40: 1527–1535.
47. Doménech-Carbó MT (2008) Novel analytical methods for characterization of binding media and protective coatings in artworks. Anal Chim Acta 621: 109–139, and references therein.
48. Colombini MP, Modugno FC (2004) Characterisation of proteinaceous binders in artistic paintings by chromatographic techniques. J Sep Sci 27: 147–160 and references therein.
49. Kaml I, Kenndler E (2007) Characterization of natural organic binding media in museum objects by capillary electrophoresis. Curr Anal Chem 3: 33–40.
50. Andreotti A, Bonaduce I, Colombini MP, Gautier G, Modugno F, Ribechini E (2006) Combined GC/MS analytical procedure for the characterization of glycerolipid, waxy, resinous, and proteinaceous materials in a unique paint microsample. Anal Chem 78: 4490–4500.
51. Learner TJS (2005) Analysis of Modern Paints. The Getty Conservation Institute, Marina del Rey.
52. Odlyha M (1995) Investigation of the binding media of paintings by thermoanalytical and spectroscopic techniques. Thermochim Acta 269/270: 705–727.
53. Goins E, Reedy ChL (2000) Digital image analysis in microscopy for objects and architectural conservation, In Objects Specialty Group Postprints, American Institut for Conservation of Historic and Artistic Works, 7, 122–137.
54. Schreiner M, Melcher M, Uhlir K (2007) Scanning electron microscopy and energy dispersive analysis: applications in the field of cultural heritage. Anal Bioanal Chem 387: 737–747.
55. Doménech-Carbó MT, Aura-Castro E, Mas-Barberá J, Martínez-Bazán ML (2006) Evaluation of morphological changes of aged acrylic, ketone and hidrocarbon resins used in contemporary artworks. Arché 1: 163–166.
56. Van den Weeerd J, van Veen MK, Heeren RMA, Boon JJ (2003) Identification of pigments in paint cross sections by reflection visible light imaging microspectroscopy. Anal Chem 75: 716–722.
57. Smith GD (2003) Infrared microspectroscopy using a synchrotron source for arts-science research. J Am Inst Conserv 42: 399–406.
58. Smith GD, Clark JH (2001) Raman microscopy in art history and conservation science. Rew Conserv 2: 92–106.
59. Giakoumaki A, Melessanaki K, Anglos D (2007) Laser-induced breakdown spectroscopy (LIBS) in archaeological science—applications and prospects. Anal Bioanal Chem 387: 749–760.

60. Wyplosz N (2003) Laser Desorption Mass Spectrometric Studies of Artists' Organic Pigments. AMOLF-FOM, Amsterdam.
61. Janssens K (2005) A survey of the recent use of X-ray beam methods for non-destructive investigations in the cultural heritage sector, Cultural Heritage Conservation and Environment Impact Assessment by Non-Destructive Testing and Micro-analysis, Van Grieken R, Janssens K (Eds) AA Balkema Publishers, London, pp. 265–308.
62. Sánchez del Río M, Gutíerrez-Léon A, Castro GR, Rubio-Zuazo J, Solís C, Sánchez-Hernández R, Robles-Camacho J, Rojas-Gaytán J (2008) Synchrotron powder diffraction on Aztec blue pigments. Appl Phys A 90: 55–60.
63. Demortier G (1991) Review of the recent applications of high energy microprobes in art and archaeology. Nucl Instrum Meth B, Beam interaction with materials and atoms 54: 334–345.
64. Geyh MA, Schleicher H (1990) Absolute Age Determination, Physical and Chemical Dating Methods and their Application. Springer, Berlin.
65. Aitken MJ (1990) Science-based Dating in Archaeology. Longman Archaeology Series, New York.
66. Martinelli N (2004) Climate from dendrochronology: latest developments and results, Glob Planet Change 40: 129–139, and references therein.
67. Tuniz C, Norton G (2008) Accelerator mass spectrometry: new trends and applications. Nucl. Instrum. Meth. B 266: 1837–1845, and references therein.
68. Wintle AG (1996) Archaeologically-relevant dating techniques for the next century small, hot and identified by acronyms. J Archaeol Sci 23: 123–138 and references therein.
69. Yip Y-Ch, Chung-wah Lam J, Tong WF (2008) Applications of lead isotope ratio measurements. Trends Anal Chem 27: 460–480 and references therein.
70. Liritzis I, Ganetsos Th (2006) Obsidian hydration dating from SIMS H+ profiling based on saturated surface (SS) layer using new software. Appl Surf Sci 252: 7144–7147.
71. Hus J, Ech-Chakrouni S, Jordanova D (2002) Origin of magnetic fabric in bricks: its implications in archaeomagnetism. Phys Chem Earth27: 1319–1331.
72. Scholz F, Nitschke L, Henrion G (1989) A new procedure for fast electrochemical analysis of solid materials. Naturwiss 76: 71–72.
73. Nitschke L, Henrion G, Damaschun F, Scholz F (1989) A new technique to study the electrochemistry of minerals. Naturwiss 76: 167–168.
74. Scholz F, Lange B (1992) Abrasive stripping voltammetry – an electrochemical solid state spectroscopy of wide applicability. Trends Anal Chem 11: 359–367.
75. Scholz F, Meyer B (1994) Electrochemical solid state analysis – state of the art. Chem Soc Rev 23: 341–347.
76. Scholz F, Meyer B (1998) Voltammetry of solid microparticles immobilized on electrode surfaces. Electroanal Chem, A Series of Advances. 20: 1–86.
77. Grygar T, Marken F, Schröder U, Scholz F (2002) Voltammetry of microparticles: a review. Coll Czech Chem Commun 67: 163–208.
78. Scholz F, Schröder U, Gulabowski R (2005) Electrochemistry of Immobilized Particles and Droplets. Springer, Berlin-Heidelberg.
79. Bard AJ, Faulkner LR (1980) Electrochemical Methods. John Wiley & Sons, New York.
80. Nicholson RS, Shain I (1964) Theory of Stationary Electrode Polarography. Anal Chem 36: 706–723.
81. Nicholson RS (1969) Some examples of the numerical solution of nonlinear integral equations. Anal Chem 37: 667–671.
82. DuVall S, McCreery RL (2000) Self-catalysis by catechols and quinones during heterogeneous electron transfer at carbon electrodes. J Am Chem Soc 122: 6759–6764.
83. Solak AO, Eichorst LR, Clark WJ, McCreery RL (2003) Modified carbon surfaces as "organic electrodes" that exhibit conductance switching. Anal Chem 75: 296–305.
84. Wang J (1985) Stripping Analysis. Principles, Instrumentation and Applications. VCH, Weinheim-New York.
85. Lovric M (2002) Square-wave voltammetry, Electroanalytical Methods, Scholz F (Ed), Springer, Berlin.

86. Bard AJ, Inzelt G, Scholz F (Eds) (2008) Electrochemical Dictionary, Springer, Berlin.
87. Wong DKY, Ewing AG (1990) Anodic stripping voltammetry at mercury films deposite on ultrasmall carbón-ring electrodes. Anal Chem 62: 2697–2702.
88. Forster RJ (1994) Microelectrodes: new dimensions in electrochemistry. Chem Soc Rev 289–297.
89. Fritz H (1929) Studien über die Empfindlichkeit einiget charakteristischer chemischer Farbreaktionen mit Hilfe del Elektro-Tüpfelmethode. Z Analyt Chem 78: 418–427.
90. Glazunov A (1929) Electrochemical reproduction of macrostructure. Chim Ind Paris Spec number 425.
91. Weisz H (1970) Microanalysis by the Ring Oven Technique, 2nd ed. Pergamon Press, Oxford.
92. Stephen WI (1956) The use of the Weisz ring oven in electrographic analysis. Mikrochim Acta 44: 1531–1539.
93. Chatterjee R, Dey AK (1967) Schematic qualitative analysis of cations by the ring oven technique. Microchem J 12: 151–156.
94. West PW, Pitombo LR (1967) Microdetermination of copper using dithiooxamide crayons and the ring-oven technique. Anal Chim Acta 37: 374–378.
95. Ghose AK, Dey AK (1970) A scheme of qualitative analysis for twenty common cations. Analyst 95: 698–701.
96. Dey, AK, Ghose AK, Shukla DK (1981) A modified ring electrographic method for the non-destructive analysis of metallic artefacts. Mikrochim Acta II: 175–181.
97. Kuwana T, French WG (1964) Electrooxidation or reduction of organic compounds into aqueous solutions using carbon paste electrode. Anal Chem 36: 241–242.
98. Schultz FA, Kuwana T (1965) Electrochemical studies of organic compounds disolved in carbón-paste electrodes. J Electroanal Chem 10: 95–103.
99. Bauer D, Gaillochet MP (1974) Etude du comportement de la pate de carbone a compose electroactif incorporé. Electrochim Acta 19: 597–606.
100. Lamache M, Bauer D (1979) Anodic oxidation of cuprous sulfide and the preparation of nonstoichiometric copper sulfide. Anal Chem 51: 1320–1322.
101. Bragé MC, Lamache M, Bauer D (1979) Contibution a l'etude des sulfures de cuivre non-stoichiometrics. Electrochim Acta 24: 25–30.
102. Brainina KhZ, Lesunova RP (1974) Use of a paste electrode for electrochemical phase analysis 1. Theoretical aspects. Formation of soluble products. Zh Anal Khim 29: 1302–1308.
103. Brainina KhZ, Vidrevich MB (1981) Stripping analysis of solids. J Electroanal Chem 121: 1– 28.
104. Ulakhovich NA, Medyantseva EP, Budnikov GK (1993) Carbon paste electrodes as chemical sensors in voltammetric analysis. Zh Anal Khim 48: 980–998.
105. Eguren M, Tascón-García ML, Vázquez-Barbado MD, Sánchez-Batanero P (1988) Study of electrochemical behavior of solid tin dioxide using a carbón paste electrode. Electrochim Acta 33, 1009–1011.
106. Lange B, Scholz F, Weuiss A, Schwedt G, Behnert J, Raezke KP (1993) Abrasive stripping voltammetry – the electrochemical alternative for pigment analysis. Internat Lab 23: 23–26.
107. Scholz F, Nitschke L, Henrion G (1989) Identification of solid materials with a new electrochemical technique – the abrasive stripping voltammetry. Fresenius Z Anal Chem 334: 56–58.
108. Doménech A, Doménech-Carbó MT, Gimeno JV, Bosch F, Saurí MC, Sánchez S (2001) Electrochemistry of iron oxide pigments (earths) from pictorial microsamples attached to graphite/polyester composite electrodes. Analyst 126: 1764–1722.
109. Kulesza PJ, Jedral T, Galkus Z (1989) A new development in polinuclear inorganic films: Silver(I)/"crosslinked" nickel(II)-hexacyanoferrate(III,II) microstructures. Electrochim Acta 34: 851–853.
110. Ghosh PK, Bard AJ (1983) Clay-modified electrodes. J Am Chem Soc 105: 5691–5693.
111. Doménech A, Doménech-Carbó MT, Moyá M, Gimeno JV, Bosch F (2000) Identification of inorganic pigments from paintings and polychromed sculptures immobilized into polymer film electrodes by stripping differential pulse voltammetry. Anal Chim Acta 407: 275–289.

112. Doménech A, Doménech-Carbó MT, Moyá M, Gimeno JV, Bosch F (2000) Voltammetric identification of lead (II) and (IV) in mediaeval glazes in abrasion-modified carbon paste and polymer film electrodes. Application to the study of alterations in archaelogical ceramic. Electroanalysis 12: 120–127.
113. Lorenzo L, Encinas P, Tascón ML, Vázquez MD, de Francisco C, Sánchez-Batanero P (1997) Electrochemical study of manganese and iron compounds at carbon paste electrodes with electrolytic binder. Application to the characterization of manganese ferrite. J Solid State Electrochem 1: 232–240.
114. Encinas-Bachiller P, Lorenzo L, Tascón-García ML, Vázquez-Barbado MD, Sánchez-Batanero P (1994) Electrochemical study of iron(II) and iron(III) compound mixtures in the solid state. Application to magnetite characterization. J Electroanal Chem 371: 161–166.
115. Lovric M, Scholz F (1997) A model for the propagation of a redox reaction through microcrystals, J Solid State Electrochem 1: 108–113.
116. Oldham KB (1998) Voltammetry at a three-phase junction. J Solid State Electrochem 2: 367–377.
117. Lovric M, Scholz F (1999) A model for the coupled transport of ions and electrons in redox conductive microcrystals, J Solid State Electrochem 3: 172–175.
118. Schröder U, Oldham KB, Myland JC, Mahon PJ, Scholz F (2000) Modelling of solid state voltammetry of immobilized microcrystals assuming an initiation of the electrochemical reaction at a three-phase junction. J Solid State Electrochem 4: 314–324.
119. Bond AM, Marken F, Hill E, Compton RG, Hügel H (1997) The electrochemical reduction of indigo dissolved in organic solvents and as a solid mechanically attached to a basal plane pyrolytic graphite electrode immersed in aqueous electrolyte solution. J Chem Soc Perkin Trans 2: 1735–1742.
120. Komorsky-Lovric S (1997) Voltammetry of azobenzene microcrystal. J Solid State Electrochem 1: 94–99.
121. Komorsky-Lovric S, Mirceski V, Scholz F (1999) Voltammetry of organic microparticles. Mikrochimica Acta 132: 67–77.
122. Doménech A, Doménech-Carbó MT, Saurí MC, Gimeno JV, Bosch F (2003) Electrochemical identification of anthraquinone-based dyes in solid microsamples by square wave voltammetry using graphite/polyester composite electrodes. Anal Bioanal Chem 375: 1169–1175.
123. Grygar T, Kuckova S, Hradil D, Hradilova D (2003) Electrochemical analysis of natural solid organic dyes and pigments. J Solid State Electrochem 7: 706–713.
124. Doménech A, Doménech-Carbó MT, Saurí MC (2005) Electrochemical identification of flavonoid dyes in work of art samples by abrasive voltammetry at paraffin-impregnated graphite electrodes. Talanta 66: 769–782.
125. Doménech A, Doménech-Carbó MT, Saurí MC, Gimeno JV, Bosch F (2005) Identification of curcuma and safflower dyes by voltammetry of microparticles using paraffin-impregnated graphite electrodes. Microchim Acta 152: 75–84.
126. Doménech A, Doménech-Carbó MT, Mas X (2007) Identification of lead pigments in nanosamples from ancient paintings and polychromed sculptures using voltammetry of nanoparticles/atomic force microscopy. Talanta 71: 1569–1579.
127. Doménech A, Doménech-Carbó MT, Gimeno JV, Bosch F, Saurí MC, Casas MJ (2001) Electrochemical analysis using charge transfer coefficient/peak potential diagrams of the alteration in copper pigments from microsamples of baroque wall paintings attached to polymer film electrodes. Fresenius J Anal Chem 369: 576–581.
128. Doménech A, Doménech-Carbó MT, Osete L (2001) Identification of manganese(IV) centers in archaeological glass using micro-sample coatings attached to polymer film electrodes. Electroanalysis 13: 927–935.
129. Doménech A, Doménech-Carbó MT, Osete L, Gimeno JV, Bosch F, Mateo R (2002) Electrochemical identification of metal ions in archaeological ceramic glazes by stripping voltammetry at graphite/polyester composite electrodes. Talanta 56: 161–174.
130. Hasse U, Scholz F (2001) In situ atomic force microscopy of the reduction of lead oxide nanocrystals immobilised on an electrode surface. Electrochem Commun 3: 429–434.

131. Komorsky-Lovric S, Lovric M, Bond AM (1992) Comparison of the square-wave stripping voltammetry of lead and mercury following their electrochemical or abrasive deposition onto a paraffin impregnated graphite electrode. Anal Chim Acta 258: 299–305.
132. Meyer B, Ziemer B, Scholz F (1995) *In Situ* X-ray diffraction study of the electrochemical reduction of tetragonal lead oxide and orthorhombic Pb(OH)Cl mechanically immobilized on a graphite electrode. J Electroanal Chem 392: 79–83.
133. Doménech A, Doménech-Carbó MT, Edwards HGM (2008) Quantitation from Tafel analysis in solid-state voltammetry. Application to the study of cobalt and copper pigments in severely damaged frescoes. Anal Chem 80: 2704–2716.
134. Burke LD, O'Sullivan JF (1992) A study of electrocatalytic behavior of gold in acid using a.c. voltammetry. Electrochim Acta 37: 2087–2094.
135. Cadle SH, Bruckenstein S (1974) Ring-disk electrode study of the anodic behavior of gold in 0.2 M sulfuric acid. Anal Chem 46: 16–20.
136. Doménech A, Doménech-Carbó MT, Osete L (2004) Electrochemistry of archaeological metals: an approach from the voltammetry of microparticles, In Trends in Electrochemistry and Corrosion at the Beginning of the 21st Century, Brillas E, Cabot PL (Eds), Universitat de Barcelona, Barcelona, pp. 857–871.
137. Grygar T, Bezdicka P, Hradil D, Doménech A, Marken F, Pikna L, Cepriá G (2002) Voltammetric analysis of iron pigments. Analyst 127: 1100–1107.
138. Grygar T, Hradilová, J, Hradil D, Bezdiéka P, Bakarddjieva S (2003) Analysis of earthy pigments in grounds of Baroque paintings. Anal Bioanal Chem 375: 1154–1160.
139. Doménech A, Doménech-Carbó MT, Edwards HGM. (2007) Identification of earth pigments in severely damaged frescoes by applying multivariate chemometric methods to solid state voltammetry. Electroanalysis 19: 1890–1900.
140. Mouhandess MT, Chassagneux F, Vittori O (1982) Electrochemical behaviour of α-iron oxide using carbon paste electrodes: influence of particle size. J Electroanal Chem.131: 367–371.
141. Mancey DS, Shoesmith DW Lipkowski J, McBride AC, Noel J (1993) An electrochemical investigation of the dissolution of magnetite in acidic electrolytes. J Eelctrochem Soc 140: 637–642.
142. Encinas-Bachiller P, Tascón-García ML, Vázquez-Barbado MD, Sánchez-Batanero P (1994) Electroanalytical study of copper and iron compounds in the solid state: application to copper ferrite characterization. J Electroanal Chem 367: 99–108.
143. Grygar T (1995) Kinetics of electrochemical reductive dissolution of iron(III) hydroxy-oxides. Collect Czech Chem Commun 60: 1261–1273.
144. Grygar T, Subrt J, Bohaceck J (1995) Electrochemical dissolution of goethite by abrasive stripping voltammetry. Collect Czech Chem Commun 6: 950–959.
145. Grygar T (1996) Electrochemical dissolution of iron(III) hydroxy-oxides: more information about the particles. Coll Czech Chem Commun 61: 93–106.
146. Grygar T (1997) Dissolution of pure and substituted goethites controlled by the surface reaction under conditions of abrasive stripping voltammetry. J Solid State Electrochem 1: 77–82.
147. Grygar T (1996) The electrochemical dissolution of iron(III) and chromium(III) oxides and ferrites under conditions of abrasive stripping voltammetry. J Electroanal Chem 405: 117–125.
148. Grygar T (1998) Phonomenological kinetics of irreversible electrochemical dissolution of metal-oxide microparticles. J Solid State Electrochem 2: 127–136.
149. Doménech A, Torres FJ, Ruiz de Sola E, Alarcón J (2006) Electrochemical detection of high oxidation states of chromium(IV and V) in chromium-doped cassiterite and tin-sphene ceramic pigmenting systems. Eur J Inorg Chem 638–648.
150. Evans DH (1990) Solution electron-transfer reactions in organic and organometallic electrochemistry. Chem Rev 90: 739–751.
151. Kano K, Uno B (1993) Surface-redox reaction mechanism of quinones adsorbed on basal-plane pyrolytic graphite electrodes. Anal Chem 58: 2009–2012.
152. Hendrickson HP, Kaufman AD, Lunte CE (1994) Electrochemistry of catechol-containing flavonoids. J Pharmac Biomed Anal 12: 325–334.

153. Ciolowski EL, Maness KM, Cahill PS, Wightman RM, Evans DH, Fosset B, Amatore C (1994) Disproportionation during electrooxidation of catecholamines at carbon-fiber microelectrodes. Anal Chem 66: 3611–3617.
154. Nagarajan P, Sulochana N, Muralidharan VS (2004) Cyclic voltammetric reduction of 7 -hydroxy and acetoxy flavones at glassy carbón electrode. Bull Electrochem 20: 93–96.
155. Brett AMO, Ghica ME (2003) Electrochemical oxidation of quercetin. Electroanalysis 15: 1745–1750.
156. Sanicanin Z, Tabakovic I (1986) Electrochemical synthetis of heterocyclic compounds. Part 16. Electrochemical transformation of 2′-hydroxychalcones into flavonoids. Tetrahedr Lett 27: 407–408.
157. Jovanovich SV, Steenken S, Hara Y, Simic MG (1996) Reduction potentials of flavonoid and modes phenoxyl radicals. Which ring in flavonoids is responsible for antioxidant activity? J Chem Soc Perkin Trans 1: 2497–2504.
158. Pournaghi-Aznar MH, Shemirani F, Pourtork S (1995) Electrochemical behavior of some naturally occurring hydroxy derivatives of 9,10-anthraquinone in chloroform at mercury and glassy carbón electrodes: application of AC polarography to the analysis of Rhubarb roots. Talanta 42: 677–684.
159. Sopchak D, Miller B, Avygal Y, Kalish R (2002) Rotating ring-disk electrode studies of the oxidation of p-methoxyphenol and hydroquinone at boron-doped diamond electrodes. J. Electroanal Chem 538: 39–45.
160. Samet Y, Abdelhedi R, Savall A (2002) A study of the electrochemical oxidation of guaiacol. Phys Chem News 8: 89–99.
161. Deffieux D, Fabre I, Courseille C, Quideau S (2002) Electrochemically induced spirolactonization of α-(methoxyphenoxy)alkonoic acids into quinone ketals. J Org Chem 67: 4458–4465.
162. Borrás, C, Rodríguez P, Laredo T, Mostany J, Scharifker BR (2004) Electrooxidation of aqueous p-methoxyphenol on lead oxide electrodes. J Appl Electrochem 34: 583–589.
163. Deffieux D, Fabre I, Titz A, Léger JM, Quideau S (2004) Electrochemical synthesis of dimerizing and nondimerizing orthoquinone. J Org Chem 69: 8731–8738.
164. Dai HP, Shin KK (1988) Voltammetric behavior of alizarin S adsorbed on electrochemically pretreated glassy carbón electrodes. Electrochim Acta 43: 2709–2715.
165. Markham KR (1982) Techniques of Flavonoid Identification. Academic Press, London, Chapter 6.
166. Lange CW, Conklin BJ, Pierpont CG (1994) Radical superexchange in semiquinone complexes containing diamagnetic metal ions. 3,6-Di-tert-butyl-1,2-semiquinoate Complexes of Zinc(II), Cobalt(III), Gallium(III), and Aluminum(III). Inorg Chem 33: 1276–1283.
167. Doménech A, Doménech-Carbó MT (2006) Chronoamperometric study of proton transfer/electron transfer in solid state electrochemistry of organic dyes. J Solid State Electrochem 10: 949–958.
168. Doménech A, Doménech-Carbó MT (2008) In situ AFM study of proton-assisted electrochemical oxidation/reduction of microparticles of organic dyes. Electrochem Commun 10: 1238–1241.
169. Doménech-Carbó MT, Casas MJ, Doménech A, Mateo R, Gimeno JV, Bosch F (2001) Analytical study of canvas painting collection from the *Basilica de la Virgen de los Desamparados* using SEM/EDX, FT-IR, GC and electrochemical techniques. Fresenius J Anal Chem 369: 571–575.
170. Sánchez S, Bosch F, Gimeno JV, Yusá DJ, Doménech A (2002) Application of XRF, XRD, thermal analysis and voltammetric techniques to the study of ancient ceramics. Anal Bioanal Chem 373: 893–900.
171. Doménech-Carbó MT, Doménech A, Osete L (2006) Analytical study of corrosion processes of archaeological glass from the Valencian Region (Spain) and its consolidation treatment. Microchim Acta 154: 123–142.

172. Edwards HGM, Doménech-Carbó MT, Hargreaves MD, Doménech A (2008) A Raman spectroscopic and combined analytical approach to the restoration of severely damaged frescoes; the palomino project. J Raman Spectrosc 39: 444–452.
173. Doménech A, Doménech-Carbó MT (2005) Electrochemical characterization of archaeological tin-opacified lead-alkali glazes and their corrosion processes. Electroanalysis 17: 1959–1969.
174. Doménech A, Doménech-Carbó MT, Ciarrocci J, Cialei V, Monteagudo A (2006) Analysis of earth pogments in Palomino's frescoes in the Santos Juanes church in Valencia (Spain) by solid-state voltammetry and FTIR spectroscopy. Arché 1: 171–176.
175. Frost RL, Ding Z, Kloprogge JT, Martens WV (2002) Thermal stability of azurite and malachite in relation to the formation of mediaeval glass and glazes. Thermochim Acta 390: 133–144.
176. Sybrandt LB, Perone SP (1972) Computerized pattern classification of strongly overlapped peaks in stationary electrode polarography. Anal Chem 44: 2331–2339.
177. Doménech A, Doménech-Carbó MT, Mas X, Ciarrocci J (2007) Simultaneous identification of lead pigments and binding media in paint samples using voltammetry of microparticles. Arché. 2: 121–124.
178. Erhardt D, Tumosa CS, Mecklenburg MF (2005) Long-term chemical and physical processes in oil paint films. Stud Conservat 50: 143–150.
179. van der Weerd J, van Loon A, Boon JJ (2005) FTIR studies of the effects of pigments on the aging of oil. Stud Conservat 50: 3–22.
180. Frost RL, Martens W, Kloprogge JT, Williams PA (2002) Raman sepctroscopy of the basic copper chloride minerals atacamite and paratacamite: implications for the study of copper, brass and bronze objects of archaeological significance. J Raman Spectrosc 33: 801–806.
181. Scott DA (2000) A review of copper chlorides and related salts in bronze corrosion and as painting pigments. Stud Conservat 45: 39–53.
182. Tennent NH, Antonio KM (1981) ICOM Committee for Conservation 6th Triennial Meeting, Ottawa.
183. Doménech A, Doménech-Carbó MT, Martínez-Lázaro I (2008) Electrochemical identification of bronze corrosion products in archaeological artefacts. A case study. Microchim Acta 162: 351–359.
184. Ramaley L, Krause MS (1969) Theory of square wave voltammetry. Anal Chem 41: 1362–1365.
185. Krause MS, Ramaley L (1969) Analytical application of square wave voltammetry. Anal Chem 41: 1365–1369.
186. O'Dea JJ, Osteryoung J, Osteryoung RA (1981) Theory of square wave voltammetry for kinetic systems. Anal Chem 53: 695–701.
187. O'Dea JJ, Osteryoung JG (1993) Characterization of quasi-reversible surface processes by square-wave voltammetry. Anal Chem 65: 3090–3097.
188. Komorsky-Lovric S, Lovric M (1995) Square-wave voltammetry of quasi-reversible surface redox reactions. J Electroanal Chem 384: 115–122.
189. Lovric M (2002) Theory of square-wave voltammetry of a reversible redox reaction complicated by the reactant adsorption. Electroanalysis 14: 405–414.
190. Gulaboski R, Mirceski V, Komorsky-Lovric S, Lovric M (2004) Square-wave voltammetry of cathodic stripping reactions. diagnostic criteria, redox kinetic measurements, and analytical applications. Electroanalysis 16: 832–842.
191. Gulaboski R, Mirceski V, Lovric M, Bogeski I (2005) Theoretical study of a surface electrode reaction preceded by a homogeneous chemical reaction under conditions of square-wave voltammetry. Electrochem Commun 7: 515–522.
192. Berrettoni M, Carpani I, Corradini N, Conti P, Fumarola G, Legnani G, Lanteri S, Marassi R, Torrelli D (2004) Coupling chemometrics and electrochemical-based sensor for detection of bacterial population. Anal Chim Acta 509: 95–101.
193. Scampicchio M, Mannino S, Zima J, Wang J (2005) Chemometrics on microchips: towards the classification of wines. Electroanalysis 17: 1215–1221.

194. Cutés A, Ibáñez AB, Céspedes F, Alegret S, del Valle M (2005) Simultaneous determination of phenolic compounds by means of an automated voltammetric "electronic tongue". Anal Bioanal Chem 382: 471–476.
195. Elias M, Chartier C, Prévot G, Garay H, Vignaud C (2006) The colour of ochres explained by their composition. Mater Sci Eng B 121: 70–80.
196. Ivanova ND, Kirillov SA, Mishchenko AB (1993) Electrochemical behaviour of non.-stoichiometric manganese oxide-hydroxide. Electrochim Acta 38: 2305–2309.
197. Amarilla JM, Tedjar F, Poisignon C (1994) Influence of KOH concentration on the g-MnO_2 redox mechanism. Electrochim Acta 39: 2321–2331.
198. Rodrigues S, Munichandraiah N, Shukla AK (1998) A cyclic voltammetric study of the kinetics and mechanism of electrodeposition of manganese dioxide. J Appl Electrochem 28: 1235–1241.
199. Bodoardo S, Brenet J, Maja M, Spinelli P (1999) Electrochemical behaviour of MnO_2 electrodes in sulphuric acid solutions. Electrochim Acta 39: 1999–2004.
200. Ratieuville Y, Wu BL, Lincot D, Vedel J, Yu LT (1999) Voltammetric and electrogravimetric study of manganese dioxide thin film electrodes. J Electrochem Soc 146: S-1–17/23.
201. Bakardjieva S, Bezdicka P, Grygar T, Vorm P (2000) Reductive dissolution of microparticulate manganese oxides. J Solid State Electrochem 4: 306–313.
202. Fetisov VB, Kozhina GA, Ermakov AN, Fetisov A, Miroshnikova EG (2007) Electrochemical dissolution of Mn_3O_4 in acid Solutions J Solid State Electrochem 11: 1205– 1210.
203. Scholz F, Hermes M (1999) The determination of the redox state of a dissolved depolariser by cyclic voltammetry in the case of electrochemically reversible systems. Electrochem Commun 1: 345–348. See corrigendum (2000) in Electrochem Commun 2: 814.
204. Doménech A, Formentín P, García H, Sabater MJ (2000) Combined electrochemical and epr studies of manganese schiff base complexes encapsulated within the cavities of zeolite Y. Eur J Inorg Chem: 1339–1344.
205. Doménech A, Sánchez S, Doménech-Carbó MT, Gimeno JV, Bosch F, Yusá DJ, Saurí MC (2002) Electrochemical determination of the Fe(III)/Fe(II) ratio in arcaheological ceramic materials using carbon paste and composite electrodes. Electroanalysis 14: 685–696.
206. Lovric M, Hermes M, Scholz F (1998) The effect of the electrolyte concentration in the solution on the voltammetric response of insertion electrodes. J Solid State Electrochem 2: 401–404.
207. Reyes-Valerio C (1993) De Bonampak al Templo Mayor: el azul Maya en Mesoamérica. Siglo XXI, México.
208. Sheppard AO (1962) Maya blue: alternative hypothesis. Am Antiq 27: 565–566.
209. Van Olphen H (1967) Maya blue: a clay mineral-organic pigment? Science 154: 465–467.
210. Kleber R, Masschelein-Kleiner L, Tissen J (1967) Study and identification of maya blue. Stud Conservat 12: 41–56.
211. José-Yacamán M, Rendón L, Arenas J, Serra Puche MC (1997) Maya blue paint: an ancient nanostructured material. Science 273: 223–224.
212. Polette LA, Meitzner G, Yacamán J, Chianelli RR (2002) Maya blue: application of XAS and HRTEM to materials science in art and archaeology. Microchem J 71: 167–174.
213. Sánchez del Río M, Martinetto P, Somogyi A, Reyes-Valerio C, Dooryhée E, Peltier N, Alianelli L, Moignard B, Pichon L, Calligaro T, Dran J-C (2004) Microanalysis study of archaeological mural samples containing maya blue pigment. Spectrochim Acta Part B: 1619–1625.
214. Sánchez del Río M, Martinetto P, Reyes-Valerio C, Dooryhée E, Suárez M (2006) Synthesis and acid resistance of maya blue pigment. Archaeometry, 48: 115–130.
215. Hubbard B, Kuang W, Moser A, Facey GA, Detellier C (2003) Structural study of maya blue: textural, thermal and solid-state multinuclear magnetic resonance characterization of the palygroskite-indigo and sepiolite-indigo adducts. Clays Clay Miner 51: 318–326.

216. Reinen D, Köhl P, Müller C (2004) The nature of the colour centres in 'maya blue' – the incorporation of organic pigment molecules into the palygorskite lattice. Zeitsch Anorg Allgem Chem 630: 97–103.
217. Ovarlez S, Chaze A-M, Giulieri F, Delamare F (2006) Indigo chemisorption in sepiolite. Application to Maya blue formation. Compt Ren Chim 9: 1243–1248.
218. Doménech A, Doménech-Carbó MT, Vázquez ML (2006) Dehydroindigo: a new piece into the maya blue puzzle from the voltammetry of microparticles approach. J Phys Chem B 110: 6027–6039.
219. Doménech A, Doménech-Carbó MT, Vázquez ML (2007) Indigo/dehydroindigo/palygorskite complex in maya blue: an electrochemical approach. J Phys Chem C 111: 4585–4595.
220. Klessinger M, Lüttke W (1963) Theoretical and spectroscopi investigation of indigo dyes II. The chromophore system of the indigo dyes. Tetrahedr Lett 19 (suppl. 2): 315–335.
221. Doménech A, Doménech-Carbó MT, Sánchez del Río M, Vázquez ML (2008) Comparative study of different indigo-clay maya blue-like systems using the voltammetry of microparticles approach. J Solid State Electrochem. In press.
222. Doménech A, Doménech-Carbó MT, Vázquez ML (2007) Chemometric study of maya blue from the voltammetry of microparticles approach. Anal Chem. 79: 2812–2821.
223. Doménech A, Doménech-Carbó MT, Vázquez ML (2007) Electrochemical monitoring maya blue preparation from maya's ancient procedures. J Solid State Electrochem 11: 1335–1346.
224. Zhang S, Meyer B, Moh GH, Scholz F (1995) Development of analytical procedures based on abrasive stripping coulometry and voltammetry for solid state phase microanalysis of natural and synthetic tin-, arsenic-, and antimony-bearing sulfosalts and sulfides of thallium, tin, lead, and silver. Electroanalysis 7: 319–328.
225. Scholz F, Müller WD, Nitschke L, Rabi F, Livanova L, Fleischfresser C, Thierfelder Ch (1990) Fast and nondestructive identification of dental alloys by abrasive stripping voltammetry. Fresenius J Anal Chem 338: 37–40.
226. Scholz F, Rabi F, Müller WD (1992) The anodic dissolution of dental amalgams studied by abrasive stripping voltammetry. Electroanalysis 4: 339–346.
227. Meyer B, Zhang S, Scholz F (1996) The quantitative analysis of mixed crystals CuSxSe1-x with abrasive stripping voltammetry and a redetermination of the solubility product of CuSe and the standard potencial of hte Cu/CuSe electrode. Fresenius' J Anal Chem 356: 267–270.
228. Scholz F, Lange B, Jaworski A, Pelzer J (1991) Analysis of powder mixtures with the help of abrasive stripping voltammetry and coulometry. Fresenius' J Anal Chem 340: 140–144.
229. Cepriá G, García-Gareta E, Pérez-Arantegui J (2005) Cadmium yellow detection and quantification by voltammetry of immobilized microparticles, Electroanalysis 17: 1078–1084.
230. Grygar T, van Oorschot IHM (2002) Voltammetric identification of pedogenic iron oxides in paleosol and loess. Electroanalysis 14: 339–344.
231. Doménech A, Doménech-Carbó MT, Osete L, Gimeno JV, Sánchez S, Bosch F (2003) Quantitation of metal ions in archaeological glass by abrasive stripping square wave voltammetry using graphite/polyester composite electrodes. Electroanalysis 15: 1465–1475.
232. Doménech A, Sánchez S, Yusá DJ, Moyá M, Gimeno JV, Bosch F (2004) Electrochemical determination of boron in minerals and ceramic materials. Anal Chim Acta 501: 103–111.
233. Doménech A, Sánchez S, Yusá DJ, Moyá M, Gimeno JV, Bosch F (2004) Determination of the boron/lead ratio in ceramic materials based on electrochemical quartz crystal microbalance. Electroanalysis 16: 1814–1822.
234. Doménech A, Moyá M, Doménech-Carbó MT (2004) Standard addition method applied to solid-state striping voltammetry. Determination of zirconium in minerals and ceramic materials. Anal Bioanal Chem 380: 146–156.
235. Saxberg BEH, Kowalski BR (1979) Generalized standard addition method. Anal Chem 51: 1031–1038.
236. Vadeginste BGM, Masart DL, Budens LMC, De Jong S, Lewi PJ, Smeyers-Verbeke J (1998) Handbook of Chemometrics and Qualimetrics, Elesvier, Amsterdam.
237. Bosch F, Campins P (1988) H-pomint standard additions method. Part 1. Fundamentals and application to analytical spectroscopy. Analyst 113: 1011–1016.

238. Campins P, Bosch F, Verdú J, Molins C (1994) Study of the behaviour of the absorbent blanks in analytical procedures by using the H-point standard additions method (HPSAM). Talanta 41: 39–52.
239. Campins P, Verdú J, Bosch F (1994) Development of the H-point standard additions method for the use of spectrofluorimetry and synchronous spectrofluorimetry. Analyst 119: 2123–2128.
240. Shams E, Abdollahi H, Yekehtaz M, Hajian R (2004) H-point standard additions method in the analysis by differential pulse anodic stripping voltammetry: Simultaneous determination of lead and thin.Talanta 63: 359–364.
241. Bosch F, Doménech A, Doménech-Carbó MT, Gimeno JV (2007) H-point standard addition method applied to voltammetry of microparticles. Quantitation of dyes in pictorial samples. Electroanalysis 19: 1575–1584.
242. Doménech A, Doménech-Carbó MT, Gimeno JV, Bosch F (2006) H-point standard addition method applied to solid state stripping voltammetry. Application to the determination of lead and tin in archaeological glazes. Anal Bioanal Chem 385: 1552–1561.
243. Lovric M, Komorsky-Lovric S (1988) Square-wave voltammetry of an adsorbed reactant. J Electroanal Chem 248: 239–253.
244. Lovric M, Komorsky-Lovric S (1989) Theory of square-wave stripping voltammetry with adsorptive accumulation. Fresenius J Anal Chem 334: 289–294.
245. Komorsky-Lovric S, Lovric M, Bond AM (1991) Theory of square-wave voltammetry of an immobilized reactant. J Electroanal Chem 319: 1–18.
246. Rehren, TH, Pernicka, E (2008) Coins, artefacts and isotopes – archaeology and archaeometry. Archeometry 50(2): 232–248.
247. Fontana MG, Greene, ND (1978) Corrosion Engineering, McGraw-Hill, New York.
248. Bertholon R (2000) La limite de la surface d'origine des objets métalliques archéologiques. Caractérisation, localisation et approche des mécanismes de conservation. PhD Thesis, Paris I – Sorbonne, Paris.
249. Scott DA (1990) Bronze desease: a review of some chemical problems and the role of relative humidity, JAIC 29: 193–206.
250. Schweizer F (1994) Objets en bronze provenant de sites lacustres: de leur patine à leur biographie, in: L'oeuvre d'art sous le regard des sciences, A. Rinuy and F. Schweizer, ed., Editions Slatkine, Genève 143–157.
251. Robbiola L, Blengino J-M, Fiaud C (1998) Morphology and mechanisms of formation of natural patinas on archaeological Cu-Sn alloys. Corros Sci 40: 2083–2111.
252. Neff D, Dillmann P, Bellot-Gurlet L, Beranger G (2005) Corrosion of iron archaeological artefacts in soil: characterisation of the corrosion system. Corros Sci 47:515–535.
253. Memet JB, Sabot R, Deslouis C, Compère C (1998) Euromat'98, Lisboa 233.
254. MacLeod ID (1991) Identification of corrosion products on non-ferrous metal artifacts recovered from shipwrecks. Stud Conserv 36: 222–234.
255. Wanhill RJH (2005) Embrittlement of ancient silver, J Fail Anal Prev 5: 33–46.
256. Rice DW, Peterson P, Rigby EB, Phipps PBP, Cappell RJ, Tremoureux R (1981) Atmospheric corrosion of copper and silver, J Electrochem Soc 128: 275–284.
257. Leygraf C, Graedel TE (2000) Atmospheric Corrosion, Electrochemical Society Series, John Wiley and Sons Inc., New York.
258. Dent Weil P (1977) A review of the history and practice of patination, NBSSP 479. Proceedings of a Seminar, Corrosion and Metal Artifacts – A Dialogue Between Conservators and Archaeologists and Corrosion Scientists held at the National Bureau of Standards, Gaithersburg, Maryland, March 17 and 18, 1976, 77–92.
259. Amarger A (2001) Les bourgeois de Calais de Rodin, in La conservation des métaux, Volfosvsky C ed. CNRS Editions 213–220.
260. Alunno-Rossetti V, Marabelli M (1976) Analyses of the patinas of a gilded horse of St. Mark's basicia in Venice: corrosion mechanisms and conservation problems. Stud Conserv 21:161–170.
261. Erhardt D, Hopwood W, Padfield T, Veloz NF (1984) The durability of Incralac: examination of a ten year old treatment, ICOM – 7th Triennial Meeting Copenhagen 84.22.1–3.

262. Costa V, Texier A (2003) Restoration of cultural heritage: stainless steel as alternative, NACE, Northern Area–Eastern Conference, Ottawa.
263. Texier A (2001) Les grands ouvrages métalliques. Le pont Alexandre III de Paris, in La conservation des métaux, C. Volfosvsky ed., CNRS Editions 185–202.
264. Oddy WA (1973) An ususpected danger in display, Mus J 73: 27–28.
265. Bradley S, Thickett D (1999) The pollution problem in perspective, Preprints from 12th Triennial Meeting, ICOM-CC, James and James, vol1: 8–13.
266. Hatchfield PB (2002) Pollutants in the museum environment, Archetype Publications, London.
267. Costa V, Urban F (2005) Lead and its alloys: metallurgy, deterioration and conservation, Rev Conserv, International Institute of Conservation 6: 48–62.
268. Costa V (2001) The deterioration of silver alloys and some aspects of their conservation. Rev Conser, International Institute of Conservation 2: 19–35
269. Richter EL, Schmidt-Ott K (1994) Zur Oberflächenbehandlung von Silber, in Metallrestaurierung, Callwey München, 182–195.
270. Sease C, Selwyn LS, Zubiate S, Bowers DF, Atkins DR (1997) Problems with coated silver: whisker formation and possible filiform corrosion, Stud Conserv 42:1–10271.
271. Green LR, Thickett D, (1991) Modern metals in museum collections, Saving the 20th century: the conservation of modern materials. Proceedings of a Conference Symposyum '91, Ottawa, Canadian Conservation Institute, 261–272.
272. Scott DA, Eggert G (2008) Iron and Steel: Corrosion, Colorants, Conservation, Archetype Publications, London.
273. Smith CS (1981) A Search for Structure: Selected Essays on Science, Art and History, MIT Press, Cambridge, 332.
274. Smith CS, Hawthorne J (1974) Mappae Clavicula: A Little Key to Medieval Techniques, American Philosophical Society, Philadelphia.
275. Pliny, Historia Naturalis, XXXIV, 149, translation K.C.Bailey, London 1932, 2: 60–61.
276. Hugues R, Rowe M (1997) The Colouring, Bronzing and Patination of Metals, Tames and Hudson Ltd., London.
277. Lechtman H (1973) Gilding of metals in pre-Columbian Peru, In Application of Science to Examination of Works of Art, W.J. Young, Ed., Boston, 38–52.
278. Beck L, Alloin E, Berthier C, Réveillon S, Costa V (2008) Silver surface enrichment controlled by simultaneous RBS for reliable PIXE analysis of ancient coins. Nucl Instrum Methods Phys Res B 266: 2320–2324.
279. Hind AM (1963) A History of Engraving and Etching from the 15th Century to the Year 1914. Dover Publications Inc., New York.
280. La Niece S, Craddock P (1993) Metal Plating and Patination: Cultural, Technical and Historical Developments, Butterworth-Heinemann, Oxford
281. Rathgen F (1924) Die Konservierung von Altertumsfunden: II und III Teil. Berlin: Walter de Gruyter.
282. Costa V (2001) The deterioration of silver alloys and some aspects of their conservation. Rev Conserv 2:19–35.
283. Plenderleith HJ, Werner AE (1971) The conservation of antiquities and works of art: treatment, repair and restoration, 2nd. edn., Oxford University Press, London.
284. Jahr U (1988) Zur Problematik der Sulfidabnahme von kunsthandwerklichen Silberobjekten. Arbeitsblätter für Restauratoren 1: 93–98
285. Lykiardopoulou-Petrou M, Beloyannis N (1998) Experimental comparative study of three different cleaning processes on Athenian silver coins (tetradrachms) from classical times'. In Look After the Pennies, Goodburn-Brown D and Jones J (Eds), Archetype Books, London, 85–88.
286. Jett PR (1993) Two examples of the treatment of ancient silver, In Current Problems in the Conservation of Metal Antiquities, 4–6 October 1993, Tokio National Research Institute of Cultural Properties, Tokio, 173–187.
287. Schiereck A (1969) Restaurierung von gut erhaltenen Eisenobjekten. Arbeitsblätter für Restauratoren 2:18–19

288. Lane H (1975) The reduction of lead, In Conservation in Archeology and the Applied Arts, IIC, London, 215–217
289. Bertholon R (2001) Nettoyage et stabilisation de la corrosion par électrolyse, In La Conservation Des Métaux, Volfovsky C (Ed) Paris: CNRS éditions 83–101
290. Carradice IA, Campbell SA (1994) The conservation of lead communion tokens by potentiostatic reduction. Stud Conserv 39:100–106.
291. Degrigny C, Jerome M, Lacoudre N (1994) Surface cleaning of silvered brass wind instruments belonging to the Sax Collection. Corros Australas 18: 16–17.
292. Degrigny C, Wery M, Vescoli M, Blengino JM (1996) Altération et nettoyage de pièces en argent doré. Stud Conserv 41: 170–178.
293. Vacquie C (1994) Mise au point d'un traitement de nettoyage de surfaces en argent doré ternies: Etude sur les filets métalliques des franges de drapeaux, Rapport D.5771 – EDF Valectra.
294. Aldaz A, España T, Montiel V, Lopez-Segura M (1986) A simple tool for the electrolytic restoration of archaeological metallic objects with localized corrosion. Stud Conserv 31: 175–6.
295. Beaudoin A, Clerice MC, Françoise J, Labbe JP, Loepper-Attia MA, Robbiola L (1997) Corrosion d'objets archéologiques en fer après déchloruration par la méthode au sulfite alcalin, in Metal 95, MacLeod ID, Pennec SL, Robbiola L (Eds), James and James, London, 170–177.
296. Bertholon R, Bell B, Blengino JM, Lacoudre N (1997) Stabilisation de la corrosion d'un objet archéologique en alliage cuivreux par électrolyse à faible polarisation dans le sesquicarbonate de sodium – dernières expériences, in Metal 95, MacLeod, I.D., Pennec, S.L., Robbiola, L. (Eds) James and James, London 209–219.
297. Degrigny C (1995) Stabilisation de moteurs d'avion immerges. Stud Conserv 40: 10–18.
298. Organ RM (1967) The reclamation of the wholly mineralized silver in the Ur lyre, In Application of Science to Examination of Works of Art, Museum of Fine Arts, Boston, 126–144.
299. Degrigny C, Le Gall R (1999) Conservation of ancient lead artifacts corroded in organic acid environments: electrolytic stabilization/consolidation. Stud Conserv 44: 157–169.
300. Gasteiger S, Eggert G (2004) How to compare reduction methods for corroded silver finds, Metal 2001: Proceedings of the International Conference on Metals Conservation, Santiago, Chile, 2–6 April 2001, Western Australian Museum 320–324.
301. Guilminot E, Baron G, Memet JB, Huet N, Le Noc E, Roze JP (2007) Electrolytic treatment of archaeological marine chloride impregnated iron objects by remote control, Metal 07, Book 3 – Use of Electrochemical Techniques in Metal Conservation, Rijksmuseum Amsterdam 38– 43.
302. Nordgreen E, Gonçalves P, Schindelholz E, Brossia CS and Yunovich M (2007) Corrosion assessement and implementation of techniques to mitigate corrosion of large artifacts from the USS Monitor (1862), Metal 07, Book 3 – Use of Electrchemical Techniques in Metal Conservation, Rijksmuseum Amsterdam 55–61.
303. Schotte B (2007) A study of the electrolytic reduction of corroded lead objects and the application, characterization and testing of a protective lead carboxylate coating, PhD dissertation Ghent University, Ghent, Belgium (https://archive.ugent.be/handle/1854/8316).
304. Tissot I, Abrantes LM (2007) Electrochemical techniques to produce and characterize protective systems on cultural silver artefacts, Metal 07, Book 3 – Use of Electrochemical Techniques in Metal Conservation, Rijksmuseum Amsterdam 62–67.
305. Costa V (2003) Museum's air quality in a tropical climate: evaluation by using metallic coupons, 5th Indoor Air Quality Meeting, Norwich (United Kingdom) http://iaq.dk/iap/iaq2003/2003_13.htm
306. Costa V, Texier A, de Reyer D (2006) Impact of environmental conditions on metallic artefacts from the treasure rooms of Reims Cathedral, Heritage, Weathering and Conservation, Fort, Alvarez de Buergo, Gomez Heras & Vazquez-Clavo (eds); Taylor & Francis Group, London 453–456.

307. Costa V, Dubus M (2007) Impact of the environmental conditions on the conservation of metal artifacts: an evaluation using electrochemical techniques, Museum Microclimates,Padfield T (Ed), The National Museum of Denmark 63–65.
308. Thickett D, Costa V (2008) The influence of dust on metal corrosion at heritage sites, 8th Indoor Air Quality Meeting, Vienna (Austria), to be published in http://www.e-preservationscience.org
309. Joiret S, Bernard MC, Costa V (2008) Assessing indoor lead corrosion using Raman spectroscopy during electrochemical reduction, 8th International Conference of Infrared and Raman user's Group, Vienna (Austria), to be published in http://www.e-preservationscience.org

About the Editor

Fritz Scholz

Fritz Scholz is a professor at the University of Greifswald in Germany. Following studies in chemistry at Humboldt University, Berlin, he obtained a Dr. rer. nat. and a Dr. sc. nat. (habilitation) from that same university. In 1987 and 1989, he worked with Alan Bond in Australia. His main interest is in electrochemistry and electroanalysis.He has published more than 250 scientific papers, and he is editor and co-author of the book "Electroanalytical Methods" (Springer 2002 and 2005; Russian Edition, BINOM 2006), coauthor of the book "Electrochemistry of Immobilized Particles and Droplets" (Springer 2005), co-editor of the "Electrochemical Dictionary" (Springer 2008), and co-editor of volumes 7a and 7b of the "Encyclopedia of Electrochemistry" (Wiley-VCH 2006). In 1997, he founded the *Journal of Solid State Electrochemistry* (Springer) and has served as Editor-in-Chief since then. He served as editor of the series "Monographs in Electrochemistry" (Springer), in which modern topics of electrochemistry are presented. Scholz introduced the technique "Voltammetry of Immobilized Microparticles" for studying the electrochemistry of solid compounds and materials, and he introduced the concept of three-phase electrodes to determine the Gibbs energies of ion transfer between immiscible liquids.

About the Authors

Antonio Doménech-Carbó

Antonio Doménech-Carbó (Valencia, 1953) holds a PhD in Chemistry (University of Valencia, 1989) and is currently a Professor in the Department of Analytical Chemistry of the University of Valencia (Spain). His research is focused on supramolecular electrochemistry, electrochemistry of porous nanostructured materials, and electroanalytical methods applied to the conservation and restoration of cultural heritage, as well as educational problems in science teaching. He has published more than 150 articles in scientific journals and monographs. He was a recipient of the "Demetrio Ribes" award (Valencian Regional Government) in 2006.

María Teresa Doménech Carbó

María Teresa Doménech Carbó studied chemistry at the University of Valencia and carried out research for her PhD at Valencia on the development of mathematical methods for suppressing interelemental effects in XRF. She took a lectureship in the science of conservation at the Universidad Politécnica de Valencia (UPV) in 1990, and a Personal Chair in 1999. She is currently the director of the Instituto Universitario de Restauración del Patrimonio in the UPV. She is interested in the development of new analytical methods in the heritage conservation context. In particular, the application of interdisciplinary approaches concerning characterization of the constitutive materials of art pieces and their conditions, alterations and biodeteriorations, recognition of damage due to previous restorations, control and assessment of conservation/restoration treatments, and the development of new materials and methods for conservation. In a research career spanning 18 years, she has published over 110 papers and nearly 40 contributions in proceedings on the chemical methods of analyzing art. She is the founder and director of *Arché* (Journal of the Heritage Conservation Institute of the UPV).

Virginia Costa

Virginia Costa was trained as an engineer in metallurgy at the Federal University of Rio Grande do Sul in Brazil (1977) and received her PhD in surface electrochemistry from the Technical University of Berlin in Germany (1992). As a professor at the Federal University in Porto Alegre (Brazil), she specializes in the corrosion and protection of metals. Since then, she has been applying her scientific and technical background in many research projects for several French and Brazilian agencies and cultural institutions, and has been acting as an expert in several international heritage projects. As a lecturer at the Institut National du Patrimoine (Paris), she has organized international workshops for conservators and published several review papers, in particular about the conservation of silver and lead artefacts, and also about the application of electrochemical techniques in the conservation of metals.

Index

MIX
Papier aus verantwortungsvollen Quellen
Paper from responsible sources
FSC® C105338

If you have any concerns about our products,
you can contact us on
ProductSafety@springernature.com

In case Publisher is established outside the EU,
the EU authorized representative is:
Springer Nature Customer Service Center GmbH
Europaplatz 3, 69115 Heidelberg, Germany

Printed by Libri Plureos GmbH
in Hamburg, Germany